Electrical and Electronic Principles 2 Checkbook

J O Bird
BSc(Hons), AFIMA, TEng(CEI), MITE

A J C May
BA, CEng, MIMechE, FITE, MBIM

Butterworths
London Boston Sydney Wellington Durban Toronto

All rights reserved. No part of this publication may be reproduced or transmitted in any form or by any means, including photocopying and recording without the written permission of the copyright holder, application for which should be addressed to the publishers. Such written permission must also be obtained before any part of this publication is stored in a retrieval system of any nature.

This book is sold subject to the Standard Conditions of Sale of Net Books and may not be resold in the UK below the net price given by the Publishers in their current price list.

First published 1981

© Butterworth & Co (Publishers) Ltd

British Library Cataloguing in Publication Data

Bird, J O
　Electrical and electronic principles 2 checkbook.
　(Butterworths checkbook series)
　1. Electric engineering
　I. Title　　II. May, A J C
　621.3　　　TK145　　　80-40734

ISBN 0-408-00635-8
ISBN 0-408-00600-5 Pbk

Typeset by Scribe Design Ltd, Gillingham, Kent
Printed and Bound by Robert Hartnoll Ltd., Bodmin, Cornwall.

Contents

Preface vii

1 Units 1
   Formulae and definitions 1
   Worked problems 3
   Further problems 5

2 DC circuit theory 8
   Formulae and definitions 8
   Worked problems 10
   Further problems 24

3 Capacitors and capacitance 32
   Formulae and definitions 32
   Worked problems 33
   Further problems 42

4 The magnetic field 48
   Formulae and definitions 48
   Worked problems 50
   Further problems 56

5 Electromagnetic induction 61
   Formulae and definitions 61
   Worked problems 64
   Further problems 73

6 Alternating voltages and currents 79
   Formulae and definitions 79
   Worked problems 84
   Further problems 95

7 Single phase AC circuits 100
   Formulae and definitions 100
   Worked problems 104
   Further problems 113

8 Semiconductor diodes 118
   Main points 118
   Worked problems 122
   Further problems 124

9 Transistors 128
   Main points 128
   Worked problems 133
   Further problems 135

10 Measuring instruments and measurements 138
   Main points 138
   Worked problems 143
   Further problems 149

Answers to multi-choice problems 152

Index 155

Note to Reader

As textbooks become more expensive, authors are often asked to reduce the number of worked and unworked problems, examples and case studies. This may reduce costs, but it can be at the expense of practical work which gives point to the theory.

Checkbooks if anything lean the other way. They let problem-solving establish and exemplify the theory contained in technician syllabuses. The Checkbook reader can gain *real* understanding through seeing problems solved and through solving problems himself.

Checkbooks do not supplant fuller textbooks, but rather supplement them with an alternative emphasis and an ample provision of worked and unworked problems. The brief outline of essential data—definitions, formulae, laws, regulations, codes of practice, standards, conventions, procedures, etc—will be a useful introduction to a course and a valuable aid to revision. Short-answer and multi-choice problems are a valuable feature of many Checkbooks, together with conventional problems and answers.

Checkbook authors are carefully selected. Most are experienced and successful technical writers; all are experts in their own subjects; but a more important qualification still is their ability to demonstrate and teach the solution of problems in their particular branch of technology, mathematics or science.

Authors, General Editors and Publishers are partners in this major low-priced series whose essence is captured by the Checkbook symbol of a question or problem 'checked' by a tick for correct solution.

Preface

This textbook of worked problems provides coverage of the Technician Education Council level 2 units in Electrical principles (syllabus U75/019) and Electrical and Electronic principles (syllabus U76/359). However it can also be regarded as a basic textbook in Electrical principles for a much wider range of courses.

The aim of the book is to provide a foundation of electrical principles needed as a basis for a range of specialisms in electrical, electronics and communications engineering and related technologies.

For syllabus U75/019,
    omit chapters 8 and 9.

For syllabus U76/359,
    omit para. 15, chapter 2,
    omit para. 17, chapter 5,
    omit paras 12, 17 and 18, chapter 6,
    omit paras 5 and 8–13, chapter 10.

Each topic considered in the text is presented in a way that assumes in the reader only the knowledge attained at TEC level 1 in Mathematics (TEC U75/005) and in Physical Science (TEC U75/004) or its équivalent.

This practical electrical principles book contains some 150 illustrations, nearly 150 detailed worked problems, followed by some 550 further problems with answers.

The authors would like to express their appreciation for the friendly co-operation and helpful advice given to them by the publishers.

Thanks are due to colleague Anthony Griffiths for helpful advice, and to Mrs Elaine Mayo for the excellent typing of the manuscript.

Finally, the authors would like to add a word of thanks to their wives, Elizabeth and Juliet, for their patience, help and encouragement during the preparation of this book.

J O Bird
A J C May
Highbury College of Technology
Portsmouth

Butterworths Technical and Scientific Checkbooks

General Editors for Science, Engineering and Mathematics titles:
J.O. Bird and A.J.C. May, Highbury College of Technology, Portsmouth.

General Editor for Building, Civil Engineering, Surveying and Architectural titles:
Colin R. Bassett, lately of Guildford County College of Technology.

A comprehensive range of Checkbooks will be available to cover the major syllabus areas of the TEC, SCOTEC and similar examining authorities. A comprehensive list is given below and classified according to levels.

Level 1 (Red covers)
Mathematics
Physical Science
Physics
Construction Drawing
Construction Technology
Microelectronic Systems
Engineering Drawing
Workshop Processes & Materials

Level 2 (Blue covers)
Mathematics
Chemistry
Physics
Building Science and Materials
Construction Technology
Electrical & Electronic Applications
Electrical & Electronic Principles
Electronics
Microelectronic Systems
Engineering Drawing
Engineering Science
Manufacturing Technology

Level 3 (Yellow covers)
Mathematics
Chemistry
Building Measurement
Construction Technology
Environmental Science
Electrical Principles
Electronics
Electrical Science
Mechanical Science
Engineering Mathematics & Science

Level 4 (Green covers)
Mathematics
Building Law
Building Services & Equipment
Construction Site Studies
Concrete Technology
Economics of the Construction Industry
Geotechnics
Engineering Instrumentation & Control

Level 5
Building Services & Equipment
Construction Technology

1 Units

A. FORMULAE AND DEFINITIONS ASSOCIATED WITH BASIC ELECTRICAL QUANTITIES

1 The system of units used in engineering and science is the **Système Internationale d'Unités** (International system of units), usually abbreviated to SI system, and is based on the metric system.

2 The **basic SI units** used in this chapter are:
   - Length — the metre (m)
   - Mass — the kilogram (kg)
   - Time — the second (s)
   - Electric current — the ampere (A)

3 **Derived SI units** use combinations of basic units and there are many of them. Two examples are:
   - Velocity — metres per second (m/s)
   - Acceleration — metres per second squared (m/s^2)

4 The **unit of charge** is the coulomb (C) where one coulomb is one ampere second. (1 coulomb = 6.24×10^{18} electrons). The coulomb is defined as the quantity of electricity which flows past a given point in an electric circuit when a current of one ampere is maintained for one second. Thus,

   charge, in coulombs $\quad Q = It$

   where I is the current in amperes and t is the time in seconds.

5 The **unit of force** is the newton (N) where one newton is one kilogram metre per second squared. The newton is defined as the force which, when applied to a mass of one kilogram, gives it an acceleration of one metre per second squared. Thus,

   Force, in newtons $\quad F = ma$

   where m is the mass in kilograms and a is the acceleration in metres per second squared. Gravitational force, or weight, is mg, where $g = 9.81$ m/s^2.

6 The **unit of work or energy** is the joule (J) where one joule is one newton metre. The joule is defined as the work done or energy transferred when a force of one newton is exerted through a distance of one metre in the direction of the force. Thus

   work done on a body, in joules $\quad W = Fs$

   where F is the force in newtons and s is the distance in metres moved by the body in the direction of the force. Energy is the capacity for doing work.

7 The **unit of power** is the watt (W) where one watt is one joule per second. Power is defined as the rate of doing work or transferring energy. Thus,

   power, in watts $\quad P = \dfrac{W}{t}$

where W is the work done or energy transferred in joules and t is the time in seconds. Thus,

energy, in joules $W = Pt$

8 Although the unit of work or energy is the joule, when dealing with large amounts of work or energy, the unit used is the kilowatt hour (kW h) where

$$\begin{aligned}1\text{ kW h} &= 1000\text{ watt hour}\\ &= 1000 \times 3600\text{ watt seconds or joules}\\ &= 3\,600\,000\text{ J}\end{aligned}$$

9 The **unit of electric potential** is the volt (V) where one volt is one joule per coulomb. One volt is defined as the difference in potential between two points in a conductor which, when carrying a current of one ampere, dissipates a power of one watt, i.e.

$$\text{volts} = \frac{\text{watts}}{\text{amperes}} = \frac{\text{joules/second}}{\text{amperes}} = \frac{\text{joules}}{\text{ampere seconds}} = \frac{\text{joules}}{\text{coulombs}}$$

A change in electric potential between two points in an electric circuit is called a potential difference. The electromotive force (emf) provided by a source of energy such as a battery or a generator is measured in volts.

10 The **unit of electric resistance** is the ohm (Ω) where one ohm is one volt per ampere. It is defined as the resistance between two points in a conductor when a constant electric potential of one volt applied at the two points produces a current flow of one ampere in the conductor. Thus,

resistance, in ohms $R = \dfrac{V}{I}$

where V is the potential difference across the two points in volts and I is the current flowing between the two points in amperes.

11 The reciprocal of resistance is called **conductance** and is measured in siemens (S).

Thus, conductance, in siemens $G = \dfrac{1}{R}$

where R is the resistance in ohms.

12 When a direct current of I amperes is flowing in an electric circuit of resistance R ohms and the voltage across the circuit is V volts, then

power, in watts $P = VI = I^2 R = \dfrac{V^2}{R}$

13 **Multiples and submultiples**

Prefix	Name	Meaning	
M	mega	multiply by 1 000 000	(i.e. $\times 10^{6}$)
k	kilo	multiply by 1 000	(i.e. $\times 10^{3}$)
m	milli	divide by 1 000	(i.e. $\times 10^{-3}$)
μ	micro	divide by 1 000 000	(i.e. $\times 10^{-6}$)
n	nano	divide by 1 000 000 000	(i.e. $\times 10^{-9}$)
p	pico	divide by 1 000 000 000 000	(i.e. $\times 10^{-12}$)

14 Terms, units and their symbols

Quantity	Quantity symbol	Unit	Unit symbol
Length	l	metre	m
Mass	m	kilogram	kg
Time	t	second	s
Velocity	v	metres per second	m/s or m s^{-1}
Acceleration	a	metres per second squared	m/s^2 or m s^{-2}
Force	F	newton	N
Electrical charge or quantity	Q	coulomb	C
Electric current	I	ampere	A
Resistance	R	ohm	Ω
Conductance	G	siemen	S
Electromotive force	E	volt	V
Potential difference	V	volt	V
Work	W	joule	J
Energy	E (or W)	joule	J
Power	P	watt	W

B. WORKED PROBLEMS INVOLVING ELECTRICAL QUANTITIES

Problem 1 A mass of 5000 g is accelerated at 2 m/s^2 by a force. Determine the force needed.

Force = mass × acceleration
$$= 5 \text{ kg} \times 2 \text{ m/s}^2 = 10 \frac{\text{kg m}}{\text{s}^2} = \textbf{10 N}$$

Problem 2 Find the force acting vertically downwards on a mass of 200 g attached to a wire. Mass = 200 g = 0.2 kg; Acceleration due to gravity, $g = 9.81$ m/s^2

Force acting downwards = weight = mass × acceleration
$$= 0.2 \text{ kg} \times 9.81 \text{ m/s}^2$$
$$= \textbf{1.962 N}$$

Problem 3 A portable machine requires a force of 200 N to move it. How much work is done if the machine is moved 20 m and what average power is utilised if the movement takes 25 s?

Work done = force × distance = 200 N × 20 m = **4000 N m or 4 kJ**

Power = $\dfrac{\text{work done}}{\text{time taken}} = \dfrac{4000}{25} \dfrac{\text{J}}{\text{s}} = \textbf{160 W}$

Problem 4 A mass of 1000 kg is raised through a height of 10 m in 20 s. What is (a) the work done and (b) the power developed?

(a) Work done = force × distance and force = mass × acceleration. Hence,

Work done = (1000 kg × 9.81 m/s^2) × (10 m)
= 98 100 N m = **98.1 kN m or 98.1 kJ**

(b) Power = $\dfrac{\text{work done}}{\text{time taken}} = \dfrac{98\,100}{20}\ \dfrac{\text{J}}{\text{s}}$ (or W) = 4905 W = **4.905 kW**

Problem 5 What current must flow if 0.24 C is to be transferred in 15 ms?

Since the quantity of electricity $Q = It$
Then
$I = \dfrac{Q}{t} = \dfrac{0.24}{15 \times 10^{-3}} = \dfrac{0.24 \times 10^3}{15} = \dfrac{240}{15} =$ **16 A**

Problem 6 If a current of 10 A flows for 4 minutes, find the quantity of electricity transferred.

Quantity of electricity $Q = It$ coulombs
$I = 10$ A; $t = 4 \times 60 = 240$ s
Hence $Q = 10 \times 240 =$ **2400 C**

Problem 7 Find the conductance of a conductor of resistance (a) 10 Ω, (b) 5 kΩ and (c) 100 mΩ.

(a) Conductance $G = \dfrac{1}{R} = \dfrac{1}{10}$ siemen = **0.1 S**

(b) $G = \dfrac{1}{R} = \dfrac{1}{5 \times 10^3}$ S $= 0.2 \times 10^{-3}$ S = **0.2 mS**

(c) $G = \dfrac{1}{R} = \dfrac{1}{100 \times 10^{-3}}$ S $= \dfrac{10^3}{100}$ S = **10 S**

Problem 8 A source emf of 15 V supplies a current of 2 A for 6 minutes. How much energy is provided in this time?

Energy = power × time and power = voltage × current. Hence
Energy = $VIt = 15 \times 2 \times (6 \times 60) = 10\,800$ W s or J

= **10.8 kJ**

Problem 9 Electrical equipment in an office takes a current of 13 A from a 240 V supply. Estimate, to the nearest pence, the cost per week of electricity if the equipment is used for 30 hours each week and 1 kW h of energy costs 4.15p.

Power = VI watts = 240 × 13 = 3120 W = 3.12 kW
Energy used per week = power × time = (3.12 kW) × (30 h)
= 93.6 kW h
Cost at 4.15p per kW h = 93.6 × 4.15 = 388.44
Hence, weekly cost of electricity = £3.88, correct to the nearest pence.

Problem 10 An electric heater consumes 3.6 MJ when connected to a 250 V supply for 40 minutes. Find the power rating of the heater and the current taken from the supply.

$$\text{Power} = \frac{\text{Energy}}{\text{time}} = \frac{3.6 \times 10^6}{40 \times 60} \frac{J}{s} \text{ (or W)} = 1500 \text{ W}$$

i.e. **Power rating of heater = 1.5 kW**

Power $P = VI$, thus $I = \frac{P}{V} = \frac{1500}{250} = 6$ A

Hence the current taken from the supply is 6 A

C. FURTHER PROBLEMS INVOLVING ELECTRICAL QUANTITIES

(a) SHORT ANSWER PROBLEMS

1. What does 'SI units' mean?
2. Complete the following: Force = ... × ....
3. What do you understand by the term 'potential difference'?
4. Define electric current in terms of charge and time.
5. Name the units used to measure (a) the quantity of electricity, (b) resistance and (c) conductance.
6. Define the coulomb.
7. Define electrical energy and name its unit.
8. Define electrical power and name its unit.
9. What is electromotive force?
10. Write down three formulae for calculating the power in a dc circuit.
11. Write down the symbols for the following quantities: (a) electrical charge; (b) work; (c) emf; (d) pd
12. State to which units the following abbreviations refer to:
    (a) A; (b) C; (c) J; (d) N; (e) m.

(b) MULTI-CHOICE PROBLEMS (Answers on page 152)

1. Which of the following formulae for electrical power is incorrect?
   (a) VI; (b) $\frac{V}{I}$; (c) I^2R; (d) $\frac{V^2}{R}$.

2. A resistance of 50 kΩ has a conductance of
   (a) 20 S; (b) 0.02 S; (c) 0.02 mS; (d) 20 kS.

3. State which of the following is incorrect;
   (a) 1 N = 1 kg m/s²; (b) 1 V = 1 J/A; (c) 30 mA = 0.03 A; (d) 1 J = 1 N/m.

4. The power dissipated by a resistor of 4 Ω when a current of 5 A passes through it is
   (a) 6.25 W; (b) 20 W; (c) 80 W; (d) 100 W.

5. 60 μs is equivalent to: (a) 0.06 s; (b) 0.000 06 s; (c) 1000 minutes: (d) 0.6 s.

6. A mass of 1200 g is accelerated at 200 cm/s² by a force. The value of the force required is:
   (a) 2.4 N; (b) 2400 N; (c) 240 kN; (d) 0.24 N.

7. A current of 3 A flows for 50 h through a 6 Ω resistor. The energy consumed by the resistor is:
   (a) 0.9 kW h; (b) 2.7 kW h; (c) 9 kW h; (d) 27 kW h.

8. What must be known in order to calculate the energy used by an electrical appliance?
   (a) voltage and current; (b) current and time of operation; (c) power and time of operation; (d) current and quantity of electricity used.

(c) CONVENTIONAL PROBLEMS

(Take $g = 9.81$ m/s² where appropriate.)

1. What force is required to give a mass of 20 kg on acceleration of 30 m/s²?

   [600 N]

2. Find the accelerating force when a car having a mass of 1.7 Mg increases its speed with a constant acceleration of 3 m/s².

   [5.1 kN]

3. A force of 40 N accelerates a mass at 5 m/s². Determine the mass.

   [8 kg]

4. Determine the force acting downwards on a mass of 1500 g suspended on a string.

   [14.72 N]

5. A force of 4 N moves an object 200 cm in the direction of the force. What amount of work is done?

   [8 J]

6. A force of 2.5 kN is required to lift a load. How much work is done if the load is lifted through 500 cm?

   [12.5 kJ]

7. An electromagnet exerts a force of 12 N and moves a soft iron armature through a distance of 1.5 cm in 40 ms. Find the power consumed.

   [4.5 W]

8. A mass of 500 kg is raised to a height of 6 m in 30 s. Find (a) the work done and (b) the power developed.

   [(a) 29.43 kN m; (b) 981 W]

9. What quantity of electricity is carried by 6.24 × 10²¹ electrons?

   [1000 C]

10. In what time would a current of 10 A transfer a charge of 50 C?

    [5 s]

11. A current of 6 A flows for 10 minutes. What charge is transferred?

[3600 C]

12. How long must a current of 100 mA flow so as to transfer a charge of 50 C?

[8 min 20 s]

13. Find the conductance of a resistor of resistance (a) 10 Ω; (b) 2 kΩ; (c) 2 mΩ.

[(a) 0.1 S; (b) 0.5 mS; (c) 500 S]

14. A conductor has a conductance of 50 μS. What is its resistance?

[20 kΩ]

15. An emf of 250 V is connected across a resistance and the current flowing through the resistance is 4 A. What is the power developed?

[1 kW]

16. 85.5 J of energy are converted into heat in 9 s. What power is dissipated?

[9.5 W]

17. A current of 4 A flows through a conductor and 10 W is dissipated. What pd exists across the ends of the conductor?

[2.5 V]

18. Find the power dissipated when:
    (a) a current of 5 mA flows through a resistance of 20 kΩ;
    (b) a voltage of 400 V is applied across a 120 kΩ resistor;
    (c) a voltage applied to a resistor is 10 kV and the current flow is 4 mA.

[(a) 0.5 W; (b) $1\frac{1}{3}$ W; (c) 40 W]

19. A battery of emf 15 V supplies a current of 2 A for 5 minutes. How much energy is supplied in this time?

[9 kJ]

20. An electric heater takes 7.5 A from a 250 V supply. Find the annual cost if the heater is used an average of 25 hours per week for 48 weeks. Assume that 1 kW h of energy costs 4p.

[£90.00]

21. A dc electric motor consumes 72 MJ when connected to a 400 V supply for 2 h 30 min. Find the power rating of the motor and the current taken from the supply.

[8 kW; 20 A]

2 DC circuit theory

A. FORMULAE AND DEFINITIONS ASSOCIATED WITH DC CIRCUIT THEORY

1. **Ohm's law** states: The current flowing in a circuit is directly proportional to the applied voltage, and inversely proportional to the resistance.

   $I = \dfrac{V}{R}$ or $V = IR$ or $R = \dfrac{V}{I}$

2. For **resistors connected in series** the equivalent resistance R_T is given by:

   $R_T = R_1 + R_2 + R_3 + \ldots + R_n$, for n resistors.

3. For **resistors connected in parallel** the equivalent resistance R_T is given by:

   $\dfrac{1}{R_T} = \dfrac{1}{R_1} + \dfrac{1}{R_2} + \dfrac{1}{R_3} + \ldots + \dfrac{1}{R_n}$, for n resistors.

4. For the special case of **two resistors connected in parallel**:

   $R_T = \dfrac{R_1 R_2}{R_1 + R_2}$ $\left(\text{i.e. } \dfrac{\text{product}}{\text{sum}}\right)$

5. (a) A network is said to be **passive** if it contains no source of emf.
   (b) A network is said to be **active** if it contains a source of emf.

6. For the series circuit shown in *Fig 1*:

   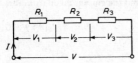
   Fig 1

   (a) $V = V_1 + V_2 + V_3$
   (b) $V_1 = IR_1$, $V_2 = IR_2$, $V_3 = IR_3$
   (c) $R_T = R_1 + R_2 + R_3$

7. The **voltage distribution** for the circuit shown in *Fig 2* is given by:

   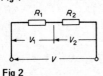
   Fig 2

   $V_1 = \left(\dfrac{R_1}{R_1 + R_2}\right) V$

   $V_2 = \left(\dfrac{R_2}{R_1 + R_2}\right) V$

8. When a continuously variable voltage is required from a fixed supply a single resistor with a sliding contact is used. Such a device is known as a **potentiometer**.

9. For the parallel circuit shown in *Fig 3*:

   (a) $I = I_1 + I_2 + I_3$ (b) $I_1 = \dfrac{V}{R_1}$, $I_2 = \dfrac{V}{R_2}$, $I_3 = \dfrac{V}{R_3}$

   (c) $\dfrac{1}{R_T} = \dfrac{1}{R_1} + \dfrac{1}{R_2} + \dfrac{1}{R_3}$

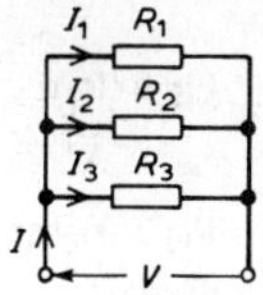

Fig 3

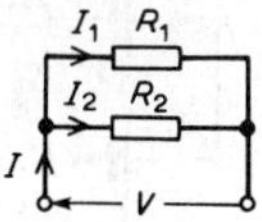

Fig 4

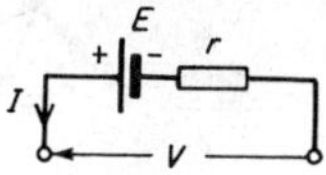

Fig 5

10 The **current division** for the circuit shown in *Fig 4* is given by:

$$I_1 = \left(\frac{R_2}{R_1+R_2}\right) I$$

$$I_2 = \left(\frac{R_1}{R_1+R_2}\right) I$$

11 For the circuit shown in *Fig 5* representing a practical source supplying energy:
$V = E - Ir$
where E is the battery emf V is the battery terminal voltage r is the internal resistance of the battery.

12 (a) When a current is flowing in the direction shown in *Fig 5* the battery is said to be **discharging** ($E > V$).
   (b) When the current flows in the opposite direction to that shown in *Fig 5* the battery is said to be **charging** ($V > E$).

13 For **cells connected in series**:
   Total emf = sum of cell's emfs
   Total internal resistance = sum of cell's internal resistances

14 For **cells connected in parallel**:
   If each cell has the same emf and internal resistance:

   Total emf = emf of one cell
   Total internal resistances of n cells = $\frac{1}{n}$ × internal resistance of one cell

15 The **superposition theorem** states:
   'In any network made up of linear resistances and containing more than one source of emf, the resultant current flowing in any branch is the algebraic sum of the currents that would flow in that branch if each source was considered separately, all other sources being replaced at that time by their respective internal resistances.'

16 **Kirchhoff's laws** state:
   (a) **Current Law**. At any junction in an electric circuit the total current flowing towards that junction is equal to the total current flowing away from the junction, i.e. $\Sigma I = 0$.
   Thus, referring to *Fig 6*: $I_1 + I_2 = I_3 + I_4 + I_5$ or $I_1 + I_2 - I_3 - I_4 - I_5 = 0$.
   (b) **Voltage Law**. In any closed loop in a network, the algebraic sum of the voltage drops (i.e. products of current and resistance) taken around the loop is equal to the resultant emf acting in that loop.

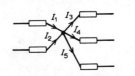
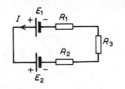

Fig 6 (left)

Fig 7 (right)

Thus, referring to *Fig 7*: $E_1 - E_2 = IR_1 + IR_2 + IR_3$.

(Note that if current flows away from the positive terminal of a source, that source is considered by convention to be positive. Thus moving anticlockwise around the loop of *Fig 7*, E_1 is positive and E_2 is negative.)

B. WORKED PROBLEMS ON DC CIRCUIT THEORY

(a) OHM'S LAW

Problem 1 A coil has a current of 50 mA flowing through it when the applied voltage is 12 V. What is the resistance of the coil?

Resistance, $R = \dfrac{V}{I} = \dfrac{12}{50 \times 10^{-3}} = \dfrac{12 \times 10^3}{50} = \dfrac{12\,000}{50} = 240\,\Omega$

Problem 2 An electric kettle has a resistance of 30 Ω. What current will flow when it is connected to a 240 V supply? Find also the power rating of the kettle.

Current $I = \dfrac{V}{R} = \dfrac{240}{30} = 8\,\text{A}$

Power $P = VI = 240 \times 8 = 1920\,\text{W} = \mathbf{1.92\,kW} = \textbf{power rating of kettle}$

Further problems on Ohm's law may be found in section C(c), problems 1 to 5, page 26.

(b) RESISTANCES IN SERIES AND IN PARALLEL

Problem 3 Find the equivalent resistance for the circuit shown in *Fig 8*.

R_3, R_4 and R_5 are connected in parallel and the equivalent resistance R is given by:

$\dfrac{1}{R} = \dfrac{1}{3} + \dfrac{1}{6} + \dfrac{1}{18} = \dfrac{6+3+1}{18} = \dfrac{10}{18}$

Hence $R = \dfrac{18}{10} = 1.8\,\Omega$

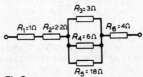

Fig 8

The circuit is now equivalent to four resistors in series and the equivalent circuit resistance = 1+2.2+1.8+4 = 9 Ω

Problem 4 Calculate the equivalent resistance between the points A and B for the circuit shown in *Fig 9*.

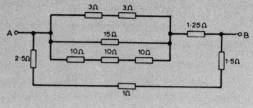

Fig 9

Combining the two 3 Ω resistors in series, the three 10 Ω resistors in series and the 2.5 Ω, 1 Ω and 1.5 Ω resistors in series gives the simplified equivalent circuit of *Fig 10*. The equivalent resistance R of 6 Ω, 15 Ω and 30 Ω in parallel is given by:

$$\frac{1}{R} = \frac{1}{6} + \frac{1}{15} + \frac{1}{30} = \frac{5+2+1}{30} = \frac{8}{30}$$

Thus $R = \frac{30}{8} = 3.75$ Ω

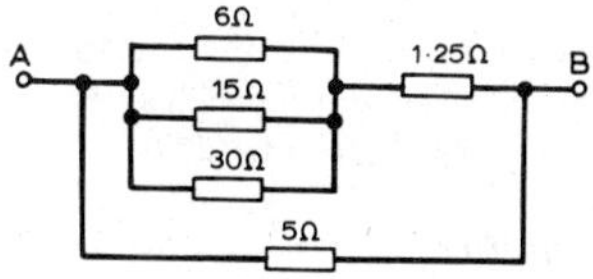

Fig 10

The equivalent circuit is now as shown in *Fig 11*. Combining the 3.75 Ω and 1.25 Ω in series gives an equivalent resistance of 5 Ω. The equivalent resistance R_T of 5 Ω in parallel with another 5 Ω resistor is given by:

$$R_T = \frac{5 \times 5}{5+5} = \frac{25}{10} = 2.5 \text{ Ω}$$

(Note that when two resistors having the same value are connected in parallel the equivalent resistance will always be half the value of one of the resistors.)

The circuit of Fig 9 can thus be replaced by a 2.5 Ω resistor placed between points A and B.

Fig 11

Problem 5 Determine the equivalent resistance for the series-parallel arrangement shown in *Fig 12*, correct to 2 decimal places.

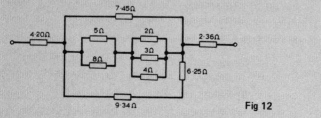

Fig 12

The equivalent resistance of 5 Ω in parallel with 8 Ω is $\frac{5\times 8}{5+8} = \frac{40}{13}$, i.e. 3.077 Ω.

The equivalent resistance R of 2 Ω, 3 Ω and 4 Ω in parallel is given by:

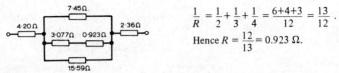

$$\frac{1}{R} = \frac{1}{2} + \frac{1}{3} + \frac{1}{4} = \frac{6+4+3}{12} = \frac{13}{12}.$$

Hence $R = \frac{12}{13} = 0.923$ Ω.

Fig 13

The equivalent resistance of 9.34 Ω and 6.25 Ω in series is 9.34+6.25 = 15.59 Ω. Thus a simplified circuit diagram is shown in *Fig 13*.

3.077 Ω in series with 0.923 Ω gives an equivalent resistance of 4.00 Ω. The equivalent resistance R_x of 7.45 Ω, 4.00 Ω and 15.59 Ω in parallel is given by:

$$\frac{1}{R_x} = \frac{1}{7.45} + \frac{1}{4.00} + \frac{1}{15.59}$$

i.e. conductance $G_x = 0.134 + 0.250 + 0.064 = 0.448$ siemens

Since $G_x = \frac{1}{R_x}$ then $R_x = \frac{1}{G_x} = \frac{1}{0.448} = 2.23$ Ω

The circuit is now equivalent to three resistors of 4.20 Ω, 2.23 Ω and 2.36 Ω connected in series, which gives an equivalent resistance of
4.20+2.23+2.36 = **8.79 Ω**.

Further problems on resistors in series and in parallel may be found in section C(c), problems 6 to 13, page 26.

(c) CURRENTS AND PD'S IN SERIES-PARALLEL CIRCUIT ARRANGEMENTS

Problem 6 Resistances of 10 Ω, 20 Ω and 30 Ω are connected (a) in series and (b) in parallel to a 240 V supply. Calculate the supply current in each case.

(a) The series circuit is shown in *Fig 14*.
The equivalent resistance

$R_T = 10\;\Omega + 20\;\Omega + 30\;\Omega = 60\;\Omega$

Supply current $I = \frac{V}{R_T} = \frac{240}{60} = \mathbf{4\ A}$

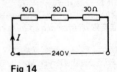

Fig 14

(b) The parallel circuit is shown in *Fig 15*. The equivalent resistance R_T of 10 Ω, 20 Ω and 30 Ω resistances connected in parallel is given by:

$$\frac{1}{R_T} = \frac{1}{10} + \frac{1}{20} + \frac{1}{30} = \frac{6+3+2}{60} = \frac{11}{60}$$

Hence $R_T = \frac{60}{11}$ Ω

Supply current $I = \frac{V}{R_T} = \frac{240}{60/11} = \frac{240 \times 11}{60} = \mathbf{44\ A}$

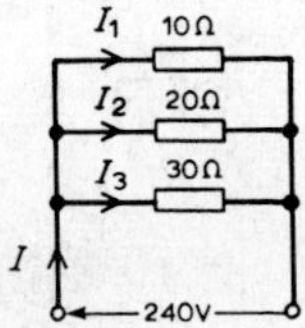

Fig 15

[Check: $I_1 = \dfrac{V}{R_1} = \dfrac{240}{10} = 24$ A

$I_2 = \dfrac{V}{R_2} = \dfrac{240}{20} = 12$ A

$I_3 = \dfrac{V}{R_3} = \dfrac{240}{30} = 8$ A

For a parallel circuit $I = I_1 + I_2 + I_3 = 24 + 12 + 8 = 44$ A, as above.]

Problem 7 For the series-parallel arrangement shown in *Fig 16* find (a) the supply current, (b) the current flowing through each resistor and (c) the pd across each resistor.

(a) The equivalent resistance R_x of R_2 and R_3 in parallel is:

$R_x = \dfrac{6 \times 2}{6+2} = \dfrac{12}{8} = 1.5 \ \Omega$

The equivalent resistance R_T of R_1, R_x and R_4 in series is:

$R_T = 2.5 + 1.5 + 4 = 8 \ \Omega$

Supply current $I = \dfrac{V}{R_T} = \dfrac{200}{8} = 25$ A

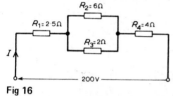

Fig 16

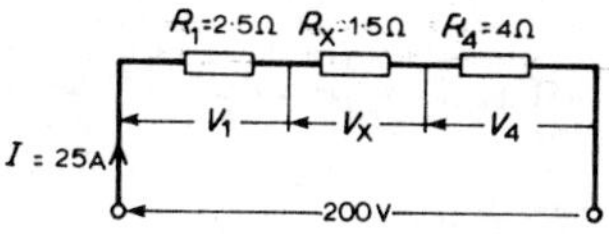

Fig 17

(b) The current flowing through R_1 and R_4 is **25 A**

The current flowing through $R_2 = \left(\dfrac{R_3}{R_2+R_3}\right)I = \left(\dfrac{2}{6+2}\right)25 = \mathbf{6.25 \ A}$

The current flowing through $R_3 = \left(\dfrac{R_2}{R_2+R_3}\right)I = \left(\dfrac{6}{6+2}\right)25 = \mathbf{18.75 \ A}$

(Note that the currents flowing through R_2 and R_3 must add up to the total current flowing into the parallel arrangement, i.e. 25 A)

(c) The equivalent circuit of *Fig 16* is shown in *Fig 17*.
pd across R_1, i.e. $V_1 = IR_1 = (25)(2.5) = \mathbf{62.5 \ V}$
pd across R_x, i.e. $V_x = IR_x = (25)(1.5) = \mathbf{37.5 \ V}$
pd across R_4, i.e. $V_4 = IR_4 = (25)(4) = \mathbf{100 \ V}$

Hence the pd across R_2 = pd across R_3 = **37.5 V**

Problem 8 For the circuit shown in *Fig 18* calculate (a) the value of resistor R_x such that the total power dissipated in the circuit is 2.5 kW, and (b) the current flowing in each of the four resistors.

13

(a) Power dissipated
$P = VI$ watts
Hence, $2500 = (250)(I)$
$$I = \frac{2500}{250} = 10 \text{ A}$$

From Ohm's law,
$R_T = \frac{V}{I} = \frac{250}{10} = 25 \text{ }\Omega$,
where R_T is the equivalent circuit resistance.

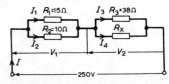

Fig 18

The equivalent resistance of R_1 and R_2 in parallel is $\frac{15 \times 10}{15+10} = \frac{150}{25} = 6 \text{ }\Omega$

The equivalent resistance of resistors R_3 and R_x in parallel is equal to $25 \text{ }\Omega - 6 \text{ }\Omega$ i.e. $19 \text{ }\Omega$.

There are three methods whereby R_x may be determined.

Method 1
The voltage $V_1 = IR$, where R is $6 \text{ }\Omega$ from above.
i.e. $V_1 = (10)(6) = 60$ V.

Hence, $V_2 = 250 \text{ V} - 60 \text{ V} = 190 \text{ V} = $ pd across $R_3 = $ pd across R_x.

$I_3 = \frac{V_2}{R_3} = \frac{190}{38} = 5$ A. Thus $I_4 = 5$ A also, since $I = 10$ A.

Thus, $R_x = \frac{V_2}{I_4} = \frac{190}{5} = 38 \text{ }\Omega$

Method 2
Since the equivalent resistance of R_3 and R_x in parallel is $19 \text{ }\Omega$

Then, $19 = \frac{38 R_x}{38 + R_x}$ $\left(\text{i.e. } \frac{\text{product}}{\text{sum}}\right)$

Hence, $19(38 + R_x) = 38 R_x$
$722 + 19 R_x = 38 R_x$
$722 = 38 R_x - 19 R_x = 19 R_x$

Thus $R_x = \frac{722}{19} = 38 \text{ }\Omega$

Method 3
When two resistors having the same value are connected in parallel the equivalent resistance is always half the value of one of the resistors. Thus, in this case, since $R_T = 19 \text{ }\Omega$ and $R_3 = 38 \text{ }\Omega$, then $R_x = 38 \text{ }\Omega$ could have been deduced on sight.

(b) Current $I_1 = \left(\frac{R_2}{R_1 + R_2}\right) I = \left(\frac{10}{15+10}\right) 10 = \left(\frac{2}{5}\right) 10 = \mathbf{4 \text{ A}}$

Current $I_2 = \left(\frac{R_1}{R_1 + R_2}\right) I = \left(\frac{15}{15+10}\right) 10 = \left(\frac{3}{5}\right) 10 = \mathbf{6 \text{ A}}$

From part (a), method 1, $I_3 = I_4 = \mathbf{5 \text{ A}}$.

Problem 9 For the arrangement shown in *Fig 19*, find the current I_x.

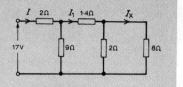

Fig 19

Commencing at the right hand side of the arrangement shown in *Fig 19*, the circuit is gradually reduced in stages as shown in *Fig 20* (a)–(d).

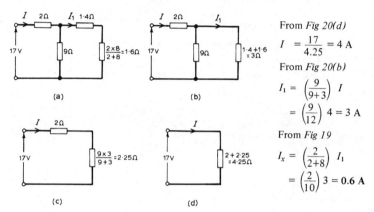

From *Fig 20(d)*
$$I = \frac{17}{4.25} = 4 \text{ A}$$

From *Fig 20(b)*
$$I_1 = \left(\frac{9}{9+3}\right) I$$
$$= \left(\frac{9}{12}\right) 4 = 3 \text{ A}$$

From *Fig 19*
$$I_x = \left(\frac{2}{2+8}\right) I_1$$
$$= \left(\frac{2}{10}\right) 3 = 0.6 \text{ A}$$

Fig 20

Further problems on currents and pd's in series-parallel arrangements may be found in section C(c), problems 14 to 20, page 27.

(d) INTERNAL RESISTANCE

Problem 10 A cell has an internal resistance of 0.03 Ω and an emf of 2.20 V. Calculate its terminal pd if it delivers (a) 1 A; (b) 10 A; (c) 40 A.

(a) For 1 A, terminal pd, $V = E - Ir = 2.20 - (1)(0.03)$ = **2.17 V**
(b) For 10 A, terminal pd, $V = E - Ir = 2.20 - (10)(0.03)$ = **1.90 V**
(c) For 40 A, terminal pd, $V = E - Ir = 2.20 - (40)(0.03)$ = **1.00 V**

Problem 11 The voltage at the terminals of a battery is 75 V when no load is connected and 72 V when a load of 60 A is connected. Find the internal resistance of the battery. What would be the terminal voltage when a load taking 40 A is connected?

When no load is connected $E = V$
Hence, the emf E of the battery is 75 V.
When a load is connected, the terminal voltage, V, is given by $V = E - Ir$
Hence $72 = 75 - (60)(r)$
$60r = 75 - 72 = 3$

$$r = \frac{3}{60} = \frac{1}{20} = 0.05 \ \Omega = \text{internal resistance of the battery}$$

When a current of 40 A is flowing then

$V = 75 - (40)(0.05)$
$ = 75 - 2 = 73 \text{ V}$

Problem 12 A battery consists of 10 cells connected in series, each cell having an emf of 2 V and an internal resistance of 0.05 Ω. The battery supplies a load R taking 4 A. Find the voltage at the battery terminals and the value of the load R.

For cells connected in series, total emf = sum of individual emfs = 20 V.
Total internal resistance = sum of individual internal resistances = 0.5 Ω
The circuit diagram is shown in *Fig 21*.

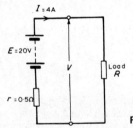

Voltage at battery terminals
$V = E - Ir$
$ = 20 - (4)(0.5)$
i.e. $V = 18 \text{ V}$

Resistance of load
$R = \dfrac{V}{I} = \dfrac{18}{4} = 4.5 \ \Omega$

Fig 21

Problem 13 Determine the equivalent resistance of the network shown in *Fig 22*. Hence determine the current taken from the supply when a battery of emf 12 V and internal resistance 0.2 Ω is connected across the terminals PQ. Find also the current flowing through the 2.9 Ω resistor and the pd across the 5.1 Ω resistor.

R_2 in series with R_3 is equivalent to 5.1 Ω + 2.9 Ω, i.e. 8 Ω.
R_1 in parallel with 8 Ω gives an equivalent resistance of $\dfrac{2 \times 8}{2 + 8} = 1.6 \ \Omega$.

1.6 Ω in series with 1.2 Ω gives an equivalent resistance of 2.8 Ω.
Hence the equivalent resistance of the network shown in Fig 22 is 2.8 Ω.
Fig 23 shows the equivalent resistance connected to the battery.

Current $I = \dfrac{E}{R_T}$, where R_T is the total circuit resistance (i.e. including the internal resistance r of the battery).

Hence $I = \dfrac{12}{2.8+0.2} = \dfrac{12}{3.0} = 4$ A

(Note that in *Fig 23* the resistances 2.8 Ω and 0.2 Ω are connected in series with each other and not in parallel.)

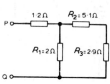

Fig 22 (above)

Fig 23 (right)

Fig 24 (extreme right)

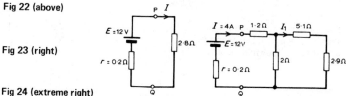

From *Fig 24*, the current flowing through the 2.9 Ω resistor, i.e. I_1 is given by
$I_1 = \left(\dfrac{2}{2+5.1+2.9}\right)(4) = 0.8$ A

The p.d. across the 5.1 Ω resistor is given by
$V = I_1(5.1) = (0.8)(5.1) = 4.08$ V

Further problems on internal resistance may be found in section C(c), Problems 21 to 27, page 28.

(e) SUPERPOSITION THEOREM

Problem 14 *Fig 25* shows a circuit containing two sources of emf, each with their internal resistances. Determine the current in each branch of the network by using the superposition theorem.

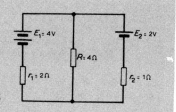

Fig 25

Procedure:
1 Redraw the original circuit with source E_2 removed, being replaced by r_2 only, as shown in *Fig 26(a)*.
2 Label the currents in each branch and their directions as shown in *Fig 26(a)* and

determine their values. (Note that the choice of current directions depends on the battery polarity, which, by convention is taken as flowing from the positive battery terminal as shown.) R in parallel with r_2 gives an equivalent resistance of

$$\frac{4 \times 1}{4+1} = 0.8 \ \Omega$$

From the equivalent circuit of *Fig 26(b)*

$$I_1 = \frac{E_1}{r_1 + 0.8} = \frac{4}{2+0.8}$$

$$= 1.429 \ A$$

From *Fig 26(a)*

$$I_2 = \left(\frac{1}{4+1}\right) I_1 = \frac{1}{5}(1.429) = 0.286 \ A$$

and

$$I_3 = \left(\frac{4}{4+1}\right) I_1 = \frac{4}{5}(1.429) = 1.143 \ A$$

3 Redraw the original circuit with source E_1 removed, being replaced by r_1 only, as shown in *Fig 27(a)*.

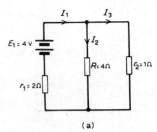

(a)

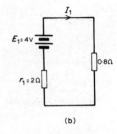

(b)

Fig 26 (above)

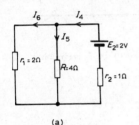

(a)

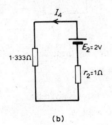

(b)

Fig 27 (left)

4 Label the currents in each branch and their directions as shown in *Fig 27(a)* and determine their values. r_1 in parallel with R gives an equivalent resistance of

$$\frac{2 \times 4}{2+4} = \frac{8}{6} = 1.333 \ \Omega$$

From the equivalent circuit of *Fig 27(b)*

$$I_4 = \frac{E_2}{1.333 + r_2} = \frac{2}{1.333+1}$$

$$= 0.857 \ A$$

From *Fig 27(a)*

$$I_5 = \left(\frac{2}{2+4}\right) I_4 = \frac{2}{6}(0.857) = 0.286 \ A$$

$$I_6 = \left(\frac{4}{2+4}\right) I_4 = \frac{4}{6}(0.857) = 0.571 \ A$$

5 Superimpose *Fig 27(a)* on to *Fig 26(a)* as shown in *Fig 28(a)*.

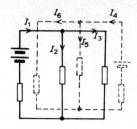

Fig 28 **Fig 29**

6 Determine the algebraic sum of the currents flowing in each branch. Resultant current flowing through source 1, i.e.

$I_1 - I_6 = 1.429 - 0.571$
$= 0.858$ A (discharging)

Resultant current flowing through source 2, i.e.

$I_4 - I_3 = 0.857 - 1.143$
$= -0.286$ A (charging)

Resultant current flowing through resistor R, i.e.

$I_2 + I_5 = 0.286 + 0.286$
$= 0.572$ A

The resultant currents with their directions are shown in *Fig 29*.

Problem 15 For the circuit shown in *Fig 30*, find, using the superposition theorem, (a) the current flowing in and the pd across the 18 Ω resistor, (b) the current in the 8 V battery and (c) the current in the 3 V battery.

---

1 Removing source E_2 gives the circuit of *Fig 31(a)*.
2 The current directions are labelled as shown in *Fig 31(a)*, I_1 flowing from the positive terminal of E_1.
From *Fig 31(b)*
$$I_1 = \frac{E_1}{3+1.8} = \frac{8}{4.8} = 1.667 \text{ A}$$

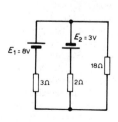

Fig 30

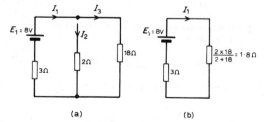

Fig 31

From *Fig 31(a)*

$$I_2 = \left(\frac{18}{2+18}\right) I_1 = \frac{18}{20} (1.667) = 1.500 \text{ A}$$

and

$$I_3 = \left(\frac{2}{2+18}\right) I_1 = \frac{2}{20} (1.667) = 0.167 \text{ A}$$

3. Removing source E_1 gives the circuit of *Fig 32(a)* (which is the same as *Fig 32(b)*).
4. The current directions are labelled as shown in *Figs 32(a)* and *32(b)*, I_4 flowing from the positive terminal of E_2.

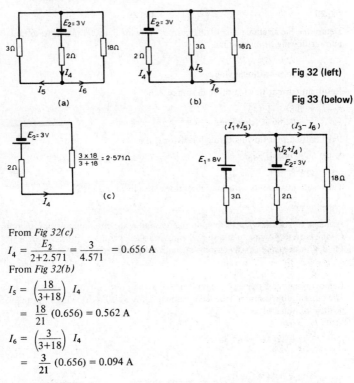

Fig 32 (left)

Fig 33 (below)

From *Fig 32(c)*

$$I_4 = \frac{E_2}{2+2.571} = \frac{3}{4.571} = 0.656 \text{ A}$$

From *Fig 32(b)*

$$I_5 = \left(\frac{18}{3+18}\right) I_4$$

$$= \frac{18}{21} (0.656) = 0.562 \text{ A}$$

$$I_6 = \left(\frac{3}{3+18}\right) I_4$$

$$= \frac{3}{21} (0.656) = 0.094 \text{ A}$$

5. Superimposing *Fig 32(a)* on to *Fig 31(a)* gives the circuit shown in *Fig 33*.
6. (a) Resultant current in the 18 Ω resistor = $I_3 - I_6$ = 0.167−0.094 = **0.073 A**
   Pd across the 18 Ω resistor = 0.073 × 18 = **1.314 V**
   (b) Resultant current in the 8 V battery = $I_1 + I_5$ = 1.667+0.562 = **2.29 A**
   (discharging)
   (c) Resultant current in the 3 V battery $I_2 + I_4$ = 1.500+0.656 = **2.156 A**
   (discharging)

Further problems on the superposition theorem may be found in section C(c), Problems 28 to 31, page 29.

(f) KIRCHHOFF'S LAWS

> *Problem 16* (a) Find the unknown currents marked in *Fig 34(a)*.
> (b) Determine the value of emf E in *Fig 34(b)*.

(a) Applying Kirchhoff's current law:

For junction B: $50 = 20 + I_1$. Hence $I_1 = 30$ A
For junction C: $20 + 15 = I_2$. Hence $I_2 = 35$ A
For junction D: $I_1 = I_3 + 120$
    i.e. $30 = I_3 + 120$. Hence $I_3 = -90$ A
(i.e. in the opposite direction to that shown in *Fig 34(a)*)
For junction E: $I_4 = 15 - I_3$
    i.e. $I_4 = 15 - (-90)$. Hence $I_4 = 105$ A
For junction F: $120 = I_5 + 40$. Hence $I_5 = 80$ A

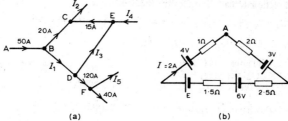

Fig 34

(b) Applying Kirchhoff's voltage law and moving clockwise around the loop of *Fig 34(b)* starting at point A:

$$3 + 6 + E - 4 = (I)(2) + (I)(2.5) + (I)(1.5) + (I)(1)$$
$$= I(2 + 2.5 + 1.5 + 1)$$
i.e. $5 + E = 2(7)$
Hence $E = 14 - 5 = 9$ V

> *Problem 17* Use Kirchhoff's laws to determine the currents flowing in each branch of the network shown in *Fig 35*.

(Note that this is the same problem as *Problem 15* and a comparison of methods may be made.)

Procedure

1 Use Kirchhoff's current law and label current directions on the original circuit diagram. The directions chosen are arbitrary, but it is usual, as a starting point, to assume that current flows from the positive terminals of the batteries. This is shown in *Fig 36* where the three branch currents are expressed in terms of I_1 and I_2 only, since the current through R is $I_1 + I_2$.

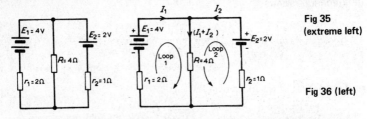

Fig 35 (extreme left)

Fig 36 (left)

2. Divide the circuit into two loops and apply Kirchhoff's voltage law to each. From loop 1 of *Fig 36*, and moving in a clockwise direction as indicated (the direction chosen does not matter), gives

$$E_1 = I_1 r_1 + (I_1 + I_2)R, \text{ i.e. } 4 = 2I_1 + 4(I_1 + I_2), \text{ i.e. } 6I_1 + 4I_2 = 4 \quad (1)$$

From loop 2 of *Fig 36*, and moving in an anticlockwise direction as indicated (once again, the choice of direction does not matter; it does not have to be in the same direction as that chosen for the first loop), gives

$$E_2 = I_2 r_2 + (I_1 + I_2)R, \text{ i.e. } 2 = I_2 + 4(I_1 + I_2), \text{ i.e. } 4I_1 + 5I_2 = 2 \quad (2)$$

3. Solve equations (1) and (2) for I_1 and I_2.

$2 \times (1)$ gives: $12I_1 + 8I_2 = 8$ (3)
$3 \times (2)$ gives: $12I_1 + 15I_2 = 6$ (4)
(3)–(4) gives: $-7I_2 = 2$ $I_2 = -\frac{2}{7} = -0.286$ A

(i.e. I_2 is flowing in the opposite direction to that shown in *Fig 36*.)

From (1) $6I_1 + 4(-0.286) = 4$
          $6I_1 = 4 + 1.144$

Hence $I_1 = \frac{5.144}{6} = 0.857$ A

Current flowing through resistance

$R = I_1 + I_2 = 0.857 + (-0.286)$
   $= 0.571$ A

(The values of the currents in each branch are seen to be the same as in *Problem 15*, when taken correct to two decimal places.) Note that a third loop is possible, as shown in *Fig 37*, giving a third equation which can be used as a check:

$E_1 - E_2 = I_1 r_1 - I_2 r_2$
$4 - 2 = 2I_1 - I_2$
$2 = 2I_1 - I_2$
[Check:
$2I_1 - I_2 = 2(0.857) - (-0.286) = 2$]

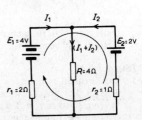

Fig 37

Problem 18 Determine, using Kirchhoff's laws, each branch current for the network shown in *Fig 38*.

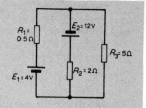

Fig 38

1. Currents, and their directions are shown labelled in *Fig 39* following Kirchhoff's current law. It is usual, although not essential, to follow conventional current flow with current flowing from the positive terminal of the source.
2. The network is divided into two loops as shown in *Fig 39*. Applying Kirchhoff's voltage law gives:

   For loop 1:

   $$E_1 + E_2 = I_1 R_1 + I_2 R_2$$
   i.e. $16 = 0.5I_1 + 2I_2$ (1)

   For loop 2:

   $$E_2 = I_2 R_2 - (I_1 - I_2)R_3$$

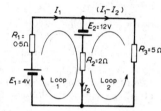

Fig 39

   Note that since loop 2 is in the opposite direction to current $(I_1 - I_2)$ the volt drop across R_3 (i.e. $(I_1 - I_2)R_3$) is by convention negative.
   Thus

   $$12 = 2I_2 - 5(I_1 - I_2) \quad \text{i.e.} \quad 12 = -5I_1 + 7I_2 \quad (2)$$

3. Solving equations (1) and (2) to find I_1 and I_2:

   $10 \times (1)$ gives $160 = 5I_1 + 20I_2$ (3)
   (2)+(3) gives $172 = 27I_2$ $I_2 = \dfrac{172}{27} = 6.37$ A

   From (1): $16 = 0.5I_1 + 2(6.37)$

   $$I_1 = \frac{16 - 2(6.37)}{0.5} = 6.52 \text{ A}$$

   Current flowing in $R_3 = I_1 - I_2 = 6.52 - 6.37 = \mathbf{0.15}$ **A**

Problem 19 For the bridge network shown in *Fig 40* determine the currents in each of the resistors.

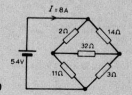

Fig 40

Let the current in the 2 Ω resistor be I_1, then by Kirchhoff's current law, the current in the 14 Ω resistor is $(I-I_1)$. Let the current in the 32 Ω resistor be I_2 as shown in *Fig 41*. Then the current in the 11 Ω resistor is (I_1-I_2) and that in the 3 Ω resistor is $(I-I_1+I_2)$. Applying Kirchhoff's voltage law to loop 1 and moving in a clockwise direction as shown in *Fig 41* gives:

$$54 = 2I_1 + 11(I_1 - I_2)$$

i.e. $\quad 13I_1 - 11I_2 = 54 \qquad (1)$

Applying Kirchhoff's voltage law to loop 2 and moving in a clockwise direction as shown in *Fig 41* gives:

$$0 = 2I_1 + 32I_2 - 14(I - I_1)$$

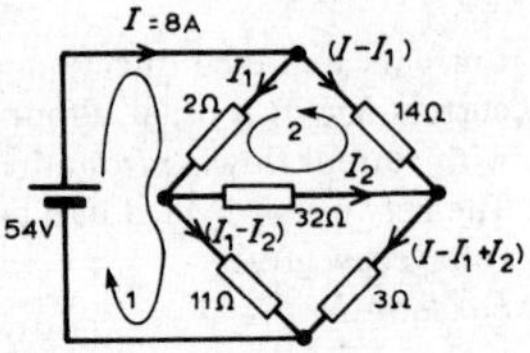

However $I = 8$ A

Fig 41

Hence $\quad 0 = 2I_1 + 32I_2 - 14(8 - I_1)$

i.e. $\quad 16I_1 + 32I_2 = 112 \qquad (2)$

Equations (1) and (2) are simultaneous equations with two unknowns, I_1 and I_2.

$16 \times (1)$ gives: $\quad 208I_1 - 176I_2 = 864 \qquad (3)$
$13 \times (2)$ gives: $\quad 208I_1 + 416I_2 = 1456 \qquad (4)$
$(4)-(3)$ gives: $\quad\quad\quad\quad\quad 592I_2 = 592$
$\quad\quad\quad\quad\quad\quad\quad\quad\quad\quad I_2 = 1$ A

Substituting for I_2 in (1) gives:

$$13I_1 - 11 = 54$$
$$I_1 = \frac{65}{13} = 5 \text{ A}$$

Hence,

the current flowing in the 2 Ω resistor $= I_1 = 5$ A
the current flowing in the 14 Ω resistor $= I - I_1 = 8 - 5 = 3$ A
the current flowing in the 32 Ω resistor $= I_2 = 1$ A
the current flowing in the 11 Ω resistor $= I_1 - I_2 = 5 - 1 = 4$ A and
the current flowing in the 3 Ω resistor $= I - I_1 + I_2 = 8 - 5 + 1 = 4$ A

Further problems on Kirchhoff's laws may be found in section C(c), Problems 32 to 38, page 30.

C. FURTHER PROBLEMS ON DC CIRCUIT THEORY

(a) SHORT ANSWER PROBLEMS

1 State Ohm's law.
2 What is a passive network?

3. What is an active network?
4. Name three characteristics of a series circuit.
5. Name three characteristics of a parallel circuit.
6. What is a potentiometer?
7. Define internal resistance and terminal pd as applied to a voltage source.
8. State Kirchhoff's current law.
9. State Kirchhoff's voltage law.
10. State, in your own words, the superposition theorem.

(b) MULTI-CHOICE PROBLEMS (Answers on page 152)

1. If two 4 Ω resistors are placed in series the effective resistance of the circuit is
   (a) 8 Ω; (b) 4 Ω; (c) 2 Ω; (d) 1 Ω.
2. If two 4 Ω resistors are placed in parallel the effective resistance of the circuit is
   (a) 8 Ω; (b) 4 Ω; (c) 2 Ω; (d) 1 Ω.
3. With the switch in *Fig 42* closed the ammeter reading will indicate
   (a) 108 A; (b) $\frac{1}{3}$ A; (c) 3 A; (d) $4\frac{3}{5}$ A.
4. A 6 Ω resistor is connected in parallel with the three resistors of *Fig 42*. With the switch closed the ammeter reading will indicate:
   (a) $\frac{3}{4}$ A; (b) 4 A; (c) $\frac{1}{4}$ A; (d) $1\frac{1}{3}$ A.
5. A 10 Ω resistor is connected in parallel with a 15 Ω resistor and the combination is connected in series with a 12 Ω resistor. The equivalent resistance of the circuit is
   (a) 37 Ω; (b) 18 Ω; (c) 27 Ω; (d) 4 Ω.

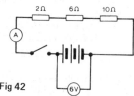

Fig 42

6. The terminal voltage of a cell of emf 2 V and internal resistance 0.1 Ω when supplying a current of 5 A will be (a) 1.5 V; (b) 2 V; (c) 1.9 V; (d) 2.5 V.
7. The effect of connecting an additional parallel load to an electrical supply source is to increase the (a) resistance of the load; (b) voltage of the source; (c) current taken from the source; (d) pd across the load.
8. The equivalent resistance when a resistor of 1/4 Ω is connected in parallel with a 1/5 Ω resistor is (a) 1/9 Ω; (b) 9 Ω.
9. Which of the following statements is true?
   For the junction in the network shown in *Fig 43*:
   (a) $I_5 - I_4 = I_3 - I_2 + I_1$
   (b) $I_1 + I_2 + I_3 = I_4 + I_5$
   (c) $I_2 + I_3 + I_5 = I_1 + I_4$
   (d) $I_1 - I_2 - I_3 - I_4 + I_5 = 0$

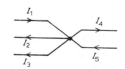

Fig 43

10. Which of the following statements is true?
    For the circuit shown in *Fig 44*:
    (a) $E_1 + E_2 + E_3 = Ir_1 + Ir_2 + Ir_3$
    (b) $E_2 + E_3 - E_1 - I(r_1 + r_2 + r_3) = 0$
    (c) $I(r_1 + r_2 + r_3) = E_1 - E_2 - E_3$
    (d) $E_2 + E_3 - E_1 = Ir_1 + Ir_2 + Ir_3$

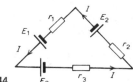

Fig 44

(c) CONVENTIONAL PROBLEMS

Ohm's law
1. Determine what voltage must be applied to a 2 kΩ resistor in order that a current of 10 mA may flow.
[20 V]

2. The hot resistance of a 240 V filament lamp is 960 Ω. Find the current taken by the lamp and its power rating.
[0.25 A; 60 W]

3. Determine the pd across a 240 Ω resistance when 12.5 mA is flowing through it.
[3 V]

4. Find the resistance of an electric fire which takes a maximum current of 13 A from a 240 V supply. Find also the power rating of the fire.
[18.46 Ω; 3.12 kW]

5. What is the resistance of a coil which draws a current of 80 mA from a 120 V supply?
[1.5 kΩ]

Resistors in series and in parallel
6. Find the equivalent resistance when the following resistances are connected (a) in series, and (b) in parallel.
 (i) 3 Ω and 2 Ω;
 (ii) 20 kΩ and 40 kΩ;
 (iii) 4 Ω, 8 Ω and 16 Ω;
 (iv) 800 Ω, 4 kΩ and 1500 Ω
[(a) (i) 5 Ω; (ii) 60 kΩ; (iii) 28 Ω; (iv) 6.3 kΩ.
 (b) (i) 1.2 Ω; (ii) $13\frac{1}{3}$ kΩ; (iii) $2\frac{2}{7}$ Ω; (iv) 461.5 Ω.]

7. If four similar lamps are connected in parallel and the total resistance of the circuit is 150 Ω, find the resistance of one lamp.
[600 Ω]

8. An electric circuit has resistances of 2.41 Ω, 3.57 Ω and 5.82 Ω connected in parallel. Find (a) the total circuit conductance, and (b) the total circuit resistance.
[(a) 0.867 S; (b) 1.154 Ω]

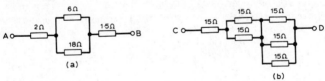

Fig 45

9. Find the total resistance between terminals A and B of the circuit shown in *Fig 45(a)*.
[8 Ω]

10. Find the equivalent resistance between terminals C and D of the circuit shown in *Fig 45(b)*.
[27.5 Ω]

11. Determine the equivalent resistance between terminals E and F of the circuit shown in *Fig 45(c)*.
[2 Ω]

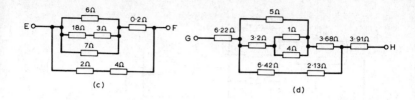

12 Find the equivalent resistance between terminals G and H of the circuit shown in *Fig 45(d)*.

[13.62 Ω]

13 State how four 1 Ω resistors must be connected to give an overall resistance of:
(a) $\frac{1}{4}$ Ω; (b) $1\frac{1}{3}$ Ω; (c) 1 Ω; (d) $2\frac{1}{2}$ Ω.

$$\begin{bmatrix} \text{(a) Four in parallel, (b) Three in parallel, in series with one} \\ \text{(c) Two in parallel, in series with another two in parallel} \\ \text{(or two in series, in parallel with another two in series)} \\ \text{(d) Two in parallel, in series with two in series.} \end{bmatrix}$$

Currents and pd's in series–parallel arrangements

14 Resistors of 20 Ω, 20 Ω and 30 Ω are connected in parallel. What resistance must be added in series with the combinations to obtain a total resistance of 10 Ω. If the complete circuit expends a power of 0.36 kW, find the total current flowing.

[2.5 Ω; 6 A]

15 (a) Calculate the current flowing in the 30 Ω resistor shown in *Fig 46*.
   (b) What additional value of resistance would have to be placed in parallel with the 20 Ω and 30 Ω resistors to change the supply current to 8 A, the supply voltage remaining constant.

[(a) 1.6 A; (b) 6 Ω]

16 For the circuit shown in *Fig 47* find (a) V_1; (b) V_2; without calculating the current flowing.

[(a) 30 V; (b) 42 V]

Fig 46 (right)

Fig 47 (extreme right)

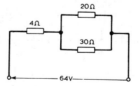
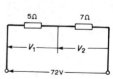

17 Determine the currents and voltages indicated in the circuit shown in *Fig 48*.

$$\begin{bmatrix} I_1 = 5 \text{ A}; I_2 = 2.5 \text{ A}; I_3 = 1\frac{2}{3}\text{A}; I_4 = \frac{5}{6}\text{A}; \\ I_5 = 3 \text{ A}; I_6 = 2 \text{ A}; V_1 = 20 \text{ V}; V_2 = 5 \text{ V}; V_3 = 6 \text{ V} \end{bmatrix}$$

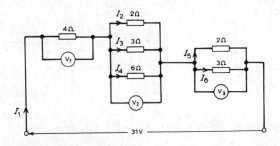

Fig 48

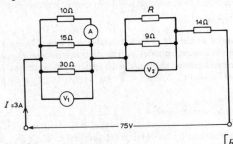

Fig 49

18 *Fig 49* shows part of an electric circuit. Find the value of resistor R and the reading on the ammeter and voltmeters.

$$\begin{bmatrix} R = 18\ \Omega;\ 1.5\ \text{A};\ V_1 = 15\ \text{V}; \\ V_2 = 18\ \text{V} \end{bmatrix}$$

19 A resistor R_x ohms is connected in series with two parallel connected resistors each of resistance 8 Ω. When the combination is connected across a 280 V supply the power taken by each of the 8 Ω resistors is 392 W. Calculate (a) the resistance of R_x, and (b) the single resistance which would take the same power as the series-parallel arrangement.

[(a) 16 Ω; (b) 20 Ω]

20 Find current I in *Fig 50*.

[1.8 A]

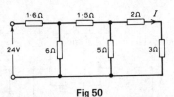

Fig 50

Internal resistance

21 A cell has an internal resistance of 0.06 Ω and an emf of 2.18 V. Find the terminal voltage if it delivers (a) 0.5 A, (b) 1 A, (c) 20 A.

[(a) 2.15 V; (b) 2.12 V; (c) 0.98 V]

22 A battery of emf 18 V and internal resistance 0.8 Ω supplies a load of 4 A. Find the voltage at the battery terminals and the resistance of the load.

[14.8 V; 3.7 Ω]

23 For the circuits shown in *Fig 51* the resistors represent the internal resistance of the batteries. Find, in each case, (a) the total emf across PQ and (b) the total equivalent internal resistances of the batteries.

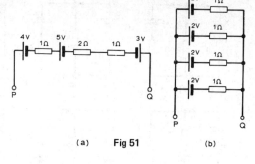

(a) **Fig 51** (b)

[(a) (i) 6 V, (ii) 2 V; (b) (i) 4 Ω, (ii) 0.25 Ω]

24 The voltage at the terminals of a battery is 52 V when no load is connected and 48.8 V when a load taking 80 A is connected. Find the internal resistance of the battery. What would be the terminal voltage when a load taking 20 A is connected?

[0.04 Ω; 51.2 V]

25 A battery of emf 36.9 V and internal resistance 0.6 Ω is connected to a circuit consisting of a resistance of 1.5 Ω in series with two resistors of 3 Ω and 6 Ω in parallel. Calculate the total current in the circuit, the current flowing through the 6 Ω resistor, the battery terminal pd and the volt drop across each resistor.

[9 A; 3 A; 31.5 V; 13.5 V; 18 V; 18 V]

26 A battery consists of four cells connected in series, each having an emf of 1.28 V and an internal resistance of 0.1 Ω. Across the terminals of the battery are two parallel resistors, $R_1 = 8$ Ω and $R_2 = 24$ Ω. Calculate the current taken by each of the resistors and the energy dissipated in the resistances, in joules, if the current flows for 3½ min.

[$I_1 = 0.6$ A; $I_2 = 0.2$ A; $W = 806.4$ J]

27 In *Fig 52*, find the total resistance measured between the points A and B. If a battery of emf 80 V and internal resistance 1 Ω is connected across AB, find the current in each resistor and the pd across R_3.

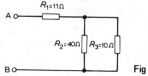

Fig 52

$$\begin{bmatrix} 19 \text{ Ω}; I_1 = 4 \text{ A}; I_2 = 0.8 \text{ A}; \\ I_3 = 3.2 \text{ A}; 32 \text{ V} \end{bmatrix}$$

Superposition theorem

28 Use the superposition theorem to find currents I_1, I_2 and I_3 of *Fig 53(a)*.

[$I_1 = 2$ A; $I_2 = 3$ A; $I_3 = 5$ A]

29 Use the superposition theorem to find the current in the 8 Ω resistor of *Fig 53(b)*.

[0.385 A]

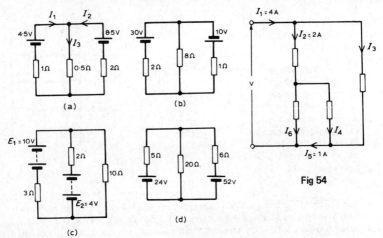

Fig 53

30 Use the superposition theorem to find the current in each branch of the network shown in *Fig 53(c)*.

$$\begin{bmatrix} \text{10 V battery discharges at 1.429 A;} \\ \text{4 V battery charges at 0.857 A;} \\ \text{Current through 10 } \Omega \text{ resistor is 0.572 A} \end{bmatrix}$$

31 Use the superposition theorem to determine the current in each branch of the arrangement shown in *Fig 53(d)*.

$$\begin{bmatrix} \text{24 V battery charges at 1.664 A;} \\ \text{52 V battery discharges at 3.280 A;} \\ \text{Current in 20 } \Omega \text{ resistor is 1.616 A.} \end{bmatrix}$$

Kirchhoff's laws

32 Find currents I_3, I_4 and I_6 in *Fig 54*.

$$[I_3 = 2 \text{ A}; I_4 = -1 \text{ A}; I_6 = 3 \text{ A}]$$

33 For the networks shown in *Fig 55*, find the values of the currents marked.

$$\begin{bmatrix} \text{(a) } I_1 = 4 \text{ A}; I_2 = -1 \text{ A}; I_3 = 13 \text{ A.} \\ \text{(b) } I_1 = 40 \text{ A}; I_2 = 60 \text{ A}; I_3 = 120 \text{ A;} \\ I_4 = 100 \text{ A}; I_5 = -80 \text{ A.} \end{bmatrix}$$

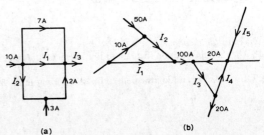

Fig 55 (a) (b)

34 Use Kirchhoff's laws to find the current flowing in the 6 Ω resistor of *Fig 56* and the power dissipated in the 4 Ω resistor.
[2.162 A; 42.07 W]

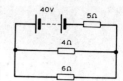

Fig 56

35 Repeat Problems 28 to 31 using Kirchhoff's laws instead of the superposition theorem.
36 Find the current flowing in the 3 Ω resistor for the network shown in *Fig 57(a)*. Find also the pd across the 10 Ω and 2 Ω resistors.

[2.715 A; 7.410 V; 3.948 V]

37 For the network shown in *Fig 57(b)* find: (a) the current in the battery; (b) the

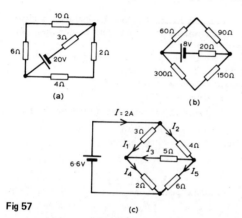

Fig 57

current in the 300 Ω resistor; (c) the current in the 90 Ω resistor; and (d) the power dissipated in the 150 Ω resistor.

$$\begin{bmatrix} \text{(a) } 60.38 \text{ mA; (b) } 15.10 \text{ mA;} \\ \text{(c) } 45.28 \text{ mA; (d) } 34.20 \text{ mW.} \end{bmatrix}$$

38 For the bridge network shown in *Fig 57(c)*, find the currents I_1 to I_5.

$$\begin{bmatrix} I_1 = 1.25 \text{ A}; I_2 = 0.75 \text{ A}; I_3 = 0.15 \text{ A}; \\ I_4 = 1.40 \text{ A}; I_5 = 0.60 \text{ A}. \end{bmatrix}$$

3 Capacitors and capacitance

A. FORMULAE AND DEFINITIONS ASSOCIATED WITH CAPACITORS AND CAPACITANCE

1. Electrostatics is the branch of electricity which is concerned with the study of electrical charges at rest. **An electrostatic field** accompanies a static charge and this is utilised in the capacitor.
2. Charged bodies attract or repel each other depending on the nature of the charge. The rule is: **like charges repel, unlike charges attract**.
3. **A capacitor** is a device capable of storing electrical energy.
4. The **charge** Q stored in a capacitor is given by:

    $Q = I \times t$ coulombs, where I is the current in amperes and t the time in seconds.

5. **A dielectric** is an insulating medium separating charged surfaces.
6. **Electric field strength, electric force, or voltage gradient,**

    $E = \dfrac{\text{pd across dielectric}}{\text{thickness of dielectric}}$, i.e. $E = \dfrac{V}{d}$ volts/m

7. **Electric flux density** $D = \dfrac{\text{electric flux}}{\text{area of one plate}} = \dfrac{\psi}{A} = \dfrac{Q}{A}$ C/m^2

    since $\psi = Q$.
    (Electric flux ψ is so defined that its value is equal to electric charge Q.)

8. **Charge Q on a capacitor** is proportional to the applied voltage V, i.e. $Q \propto V$.
9. $Q = CV$ or $C = Q/V$, where the constant of proportionality C is the **capacitance**.
10. The **unit of capacitance** is the farad F (or more usually μF$=10^{-6}$ F or pF$=10^{-12}$ F), which is defined as the capacitance of a capacitor when a pd of one volt appears across the plates when charged with one coulomb.
11. Every system of electrical conductors possesses capacitance. For example, there is capacitance between the conductor of overhead transmission lines and also between the wires of a telephone cable. In these examples the capacitance is undesirable but has to be accepted, minimised or compensated for. There are other situations, such as in capacitors, where capacitance is a desirable property.
12. The ratio of electric flux density, D, to electric field strength, E, is called **absolute permittivity**, ϵ, of a dielectric.

    Thus $\dfrac{D}{E} = \epsilon$.

13. The **permittivity of free space** is a constant, given by $\epsilon_0 = 8.85 \times 10^{-12}$ F/m.
14. **Relative permittivity**

    $\epsilon_r = \dfrac{\text{flux density of the field in the dielectric}}{\text{flux density of the field in vacuum}}$ (ϵ_r has no units.)

Examples of the values of ϵ_r include: air = 1, polythene = 2.3, mica = 3–7, glass = 5–10, ceramics = 6–1000.

15 **Absolute permittivity**, $\epsilon = \epsilon_0 \epsilon_r$.
Thus $D/E = \epsilon_0 \epsilon_r$.

16 For a **parallel plate capacitor**, capacitance is proportional to area A, inversely proportional to the plate spacing (or dielectric thickness), d, and depends on the nature of the dielectric and the number of plates, n.

Capacitance $C = \dfrac{\epsilon_0 \epsilon_r A (n-1)}{d}$ F

17 For n **capacitors connected in parallel**, the equivalent capacitance C_T is given by:

$C_T = C_1 + C_2 + C_3 + \ldots + C_n$ (similar to resistors connected in series)

Also total charge,

$Q_T = Q_1 + Q_2 + Q_3 + \ldots + Q_n$

18 For n **capacitors connected in series**, the equivalent capacitance C_T is given by:

$\dfrac{1}{C_T} = \dfrac{1}{C_1} + \dfrac{1}{C_2} + \dfrac{1}{C_3} + \ldots + \dfrac{1}{C_n}$ (similar to resistors connected in parallel)

The charge on each capacitor is the same when connected in series.

19 The maximum amount of field strength that a dielectric can withstand is called the **dielectric strength** of the material.

Dielectric strength, $E_{MAX} = \dfrac{V_{MAX}}{d}$ and $V_{MAX} = d \times E_{MAX}$.

20 The **energy**, W, **stored by a capacitor** is given by $W = \tfrac{1}{2} CV^2$ joules.

21 Practical **types of capacitor** are characterised by the material used for their dielectric. The main types include: variable air, mica, paper, ceramic, plastic and electrolytic (see *Problem 19*).

22 When a capacitor has been disconnected from the supply it may still be charged and it may retain this charge for some considerable time. Thus precautions must be taken to ensure that the capacitor is automatically discharged after the supply is switched off. This is done by connecting a high value resistor across the capacitor terminals.

B. WORKED PROBLEMS ON CAPACITORS AND CAPACITANCE

(a) ELECTRIC FLUX DENSITY AND ELECTRIC FIELD STRENGTH PROBLEMS

Problem 1 Two parallel rectangular plates measuring 20 cm by 40 cm carry an electric charge of 0.2 µC. Calculate the electric flux density. If the plates are spaced 5 mm apart and the voltage between them is 0.25 kV determine the electric field strength.

Charge $Q = 0.2$ μC $= 0.2 \times 10^{-6}$ C;

Area $A = 20$ cm $\times 40$ cm $= 800$ cm$^2 = 800 \times 10^{-4}$ m^2

Electric flux density $D = \dfrac{Q}{A} = \dfrac{0.2 \times 10^{-6}}{800 \times 10^{-4}} = \dfrac{0.2 \times 10^{4}}{800 \times 10^{6}} = \dfrac{2000}{800} \times 10^{-6}$

$$= 2.5 \text{ μC/m}^2$$

Voltage $V = 0.25$ kV $= 250$ V; Plate spacing, $d = 5$ mm $= 5 \times 10^{-3}$ m.

Electric field strength $E = \dfrac{V}{d} = \dfrac{250}{5 \times 10^{-3}} =$ **50 kV/m**

Problem 2 The flux density between two plates separated by mica of relative permittivity 5 is 2 μC/m^2. Find the voltage gradient between the plates.

Flux density $D = 2$ μC/m$^2 = 2 \times 10^{-6}$ C/m^2; $\epsilon_0 = 8.85 \times 10^{-12}$ F/m; $\epsilon_r = 5$.

$$\dfrac{D}{E} = \epsilon_0 \epsilon_r$$

Hence voltage gradient $E = \dfrac{D}{\epsilon_0 \epsilon_r}$

$$= \dfrac{2 \times 10^{-6}}{8.85 \times 10^{-12} \times 5} \text{ V/m} = \textbf{45.2 kV/m}$$

Problem 3 Two parallel plates having a pd of 200 V between them are spaced 0.8 mm apart. What is the electric field strength? Find also the flux density when the dielectric between the plates is (a) air, and (b) polythene of relative permittivity 2.3.

Electric field strength $E = \dfrac{V}{d} = \dfrac{200}{0.8 \times 10^{-3}} = 250$ kV/m

(a) For air: $\epsilon_r = 1$
$\dfrac{D}{E} = \epsilon_0 \epsilon_r$
Hence electric flux density $D = E\epsilon_0 \epsilon_r$
$= 250 \times 10^{3} \times 8.85 \times 10^{-12} \times 1$ C/m^2
$= \textbf{2.213 μC/m}^2$

(b) For polythene: $\epsilon_r = 2.3$.
Electric flux density $D = E\epsilon_0 \epsilon_r = 250 \times 10^{3} \times 8.85 \times 10^{-12} \times 2.3$ C/m^2
$= \textbf{5.089 μC/m}^2$

Further problems on electric flux density and electric field strength may be found in section C(c), Problems 1 to 6, page 44.

(b) $Q = CV$ PROBLEMS

Problem 4 (a) Determine the pd across a 4 μF capacitor when charged with 5 mC. (b) Find the charge on a 50 pF capacitor when the voltage applied to it is 2 kV.

(a) $C = 4$ μF $= 4 \times 10^{-6}$ F; $Q = 5$ mC $= 5 \times 10^{-3}$ C

Since $C = \dfrac{Q}{V}$ then $V = \dfrac{Q}{C} = \dfrac{5 \times 10^{-3}}{4 \times 10^{-6}} = \dfrac{5 \times 10^{6}}{4 \times 10^{3}} = \dfrac{5000}{4}$

Hence pd = 1250 V or 1.25 kV

(b) $C = 50$ pF $= 50 \times 10^{-12}$ F; $V = 2$ kV $= 2000$ V

$Q = CV = 50 \times 10^{-12} \times 2000 = \dfrac{5 \times 2}{10^{8}} = 0.1 \times 10^{-6}$

Hence charge = 0.1 μC

Problem 5 A direct current of 4 A flows into a previously uncharged 20 μF capacitor for 3 ms. Determine the pd between the plates.

$I = 4$ A; $C = 20$ μF $= 20 \times 10^{-6}$ F; $t = 3$ ms $= 3 \times 10^{-3}$ s

$Q = It = 4 \times 3 \times 10^{-3}$ C

$V = \dfrac{Q}{C} = \dfrac{4 \times 3 \times 10^{-3}}{20 \times 10^{-6}} = \dfrac{12 \times 10^{6}}{20 \times 10^{3}} = 0.6 \times 10^{3} = 600$ V

Hence, the pd between the plates is 600 V

Problem 6 A 5 μF capacitor is charged so that the pd between its plates is 800 V. Calculate how long the capacitor can provide an average discharge current of 2 mA.

$C = 5$ μF $= 5 \times 10^{-6}$ F; $V = 800$ V; $I = 2$ mA $= 2 \times 10^{-3}$ A

$Q = CV = 5 \times 10^{-6} \times 800 = 4 \times 10^{-3}$ C

Also,

$Q = It$ Thus, $t = \dfrac{Q}{I} = \dfrac{4 \times 10^{-3}}{2 \times 10^{-3}} = 2$ s

Hence the capacitor can provide an average discharge current of 2 mA for 2 s

Further problems on $Q = CV$ may be found in section C(c), Problems 7 to 13, page 44.

(c) PARALLEL PLATE CAPACITOR PROBLEMS

Problem 7 (a) A ceramic capacitor has an effective plate area of 4 cm² separated by 0.1 mm of ceramic of relative permittivity 100. Calculate the capacitance of the capacitor in picofarads.
(b) If the capacitor in part (a) is given a charge of 1.2 μC what will be the pd between the plates?

(a) Area $A = 4 \text{ cm}^2 = 4 \times 10^{-4} \text{ m}^2$; $d = 0.1 \text{ mm} = 0.1 \times 10^{-3}$ m;
$\epsilon_0 = 8.95 \times 10^{-12}$ F/m; $\epsilon_r = 100$.

Capacitance $C = \dfrac{\epsilon_0 \epsilon_r A}{d}$ farads $= \dfrac{8.85 \times 10^{-12} \times 100 \times 4 \times 10^{-4}}{0.1 \times 10^{-3}}$ F

$= \dfrac{8.85 \times 4}{10^{10}}$ F $= \dfrac{8.85 \times 4 \times 10^{12}}{10^{10}}$ pF = **3540 pF**

(b) $Q = CV$ thus $V = \dfrac{Q}{C} = \dfrac{1.2 \times 10^{-6}}{3540 \times 10^{-12}}$ V = **339 V**

Problem 8 A waxed paper capacitor has two parallel plates, each of effective area 800 cm². If the capacitance of the capacitor is 4425 pF determine the effective thickness of the paper if its relative permittivity is 2.5.

$A = 800 \text{ cm}^2 = 800 \times 10^{-4} \text{ m}^2 = 0.08 \text{ m}^2$; $C = 4425 \text{ pF} = 4425 \times 10^{-12}$ F;
$\epsilon_0 = 8.85 \times 10^{-12}$ F/m; $\epsilon_r = 2.5$.

Since $C = \dfrac{\epsilon_0 \epsilon_r A}{d}$

Then $d = \dfrac{\epsilon_0 \epsilon_r A}{C}$

Hence, $d = \dfrac{8.85 \times 10^{-12} \times 2.5 \times 0.08}{4425 \times 10^{-12}} = 0.0004$ m

Hence the thickness of the paper is 0.4 mm

Problem 9 A parallel plate capacitor has nineteen interleaved plates each 75 mm by 75 mm separated by mica sheets 0.2 mm thick. Assuming the relative permittivity of the mica is 5, calculate the capacitance of the capacitor.

$n = 19$; $n-1 = 18$; $A = 75 \times 75 = 5625 \text{ mm}^2 = 5625 \times 10^{-6} \text{ m}^2$;
$\epsilon_r = 5$; $\epsilon_0 = 8.85 \times 10^{-12}$ F/m; $d = 0.2 \text{ mm} = 0.2 \times 10^{-3}$ m.

Capacitance $C = \dfrac{\epsilon_0 \epsilon_r A(n-1)}{d} = \dfrac{8.85 \times 10^{-12} \times 5 \times 5625 \times 10^{-6} \times 18}{0.2 \times 10^{-3}}$ F

= **0.0224 μF or 22.4 nF**

Problem 10 A capacitor is to be constructed so that its capacitance is 0.2 μF and to take a pd of 1.25 kV across its terminals. The dielectric is to be mica which, after allowing a safety factor, has a dielectric strength of 50 MV/m. Find (a) the thickness of the mica needed and (b) the area of a plate assuming a two plate construction. (Assume ϵ_r for mica to be 6.)

(a) Dielectric strength $E = \dfrac{V}{d}$ i.e. $d = \dfrac{V}{E} = \dfrac{1.25 \times 10^3}{50 \times 10^6}$ m = 0.025 mm

(b) Capacitance $C = \dfrac{\epsilon_0 \epsilon_r A}{d}$

Hence area $A = \dfrac{Cd}{\epsilon_0 \epsilon_r} = \dfrac{0.2 \times 10^{-6} \times 0.025 \times 10^{-3}}{8.85 \times 10^{-12} \times 6}$ m²

= 0.094 16 m² = **941.6 cm²**

Further problems on parallel plate capacitors may be found in section C(c), Problems 14 to 22, page 45.

(d) PROBLEMS ON CAPACITORS CONNECTED IN PARALLEL AND IN SERIES

Problem 11 Calculate the equivalent capacitance of two capacitors of 6 μF and 4 μF connected (a) in parallel and (b) in series.

(a) In parallel, equivalent capacitance $C = C_1 + C_2 = 6$ μF + 4 μF = **10 μF**.

(b) In series, equivalent capacitance C is given by: $\dfrac{1}{C} = \dfrac{1}{C_1} + \dfrac{1}{C_2} = \dfrac{C_2 + C_1}{C_1 C_2}$

i.e. $C = \dfrac{C_1 C_2}{C_1 + C_2}$ i.e. $\dfrac{\text{product}}{\text{sum}}$

This formula is used for the special case of **two** capacitors in series (which is similar to two resistors in parallel).

Thus $C = \dfrac{6 \times 4}{6 + 4} = \dfrac{24}{10} =$ **2.4 μF**

Problem 12 What capacitance must be connected in series with a 30 μF capacitor for the equivalent capacitance to be 12 μF?

Let $C = 12$ μF (the equivalent capacitance), $C_1 = 30$ μF and C_2 be the unknown capacitance.

For two capacitors in series $\dfrac{1}{C} = \dfrac{1}{C_1} + \dfrac{1}{C_2}$

Hence $\dfrac{1}{C_2} = \dfrac{1}{C} - \dfrac{1}{C_1} = \dfrac{C_1 - C}{C C_1}$

$C_2 = \dfrac{C C_1}{C_1 - C} = \dfrac{12 \times 30}{30 - 12} = \dfrac{360}{18} =$ **20 μF**

Problem 13 Capacitances of 1 µF, 3 µF, 5 µF and 6 µF are connected in parallel to a direct voltage supply of 100 V. Determine (a) the equivalent circuit capacitance, (b) the total charge and (c) the charge on each capacitor.

(a) The equivalent capacitance C for 4 capacitors in parallel is given by:
$$C = C_1 + C_2 + C_3 + C_4$$
i.e. $C = 1+3+5+6 = 15\ \mu F$

(b) Total charge $Q_T = CV$ where C is the equivalent circuit capacitance.
i.e. $Q_T = 15 \times 10^{-6} \times 100 = 1.5 \times 10^{-3}$ C = **1.5 mC**

(c) The charge on the 1 µF capacitor $Q_1 = C_1 V = 1 \times 10^{-6} \times 100 = $ **0.1 mC**
The charge on the 3 µF capacitor $Q_2 = C_2 V = 3 \times 10^{-6} \times 100 = $ **0.3 mC**
The charge on the 5 µF capacitor $Q_3 = C_3 V = 5 \times 10^{-6} \times 100 = $ **0.5 mC**
The charge on the 6 µF capacitor $Q_4 = C_4 V = 6 \times 10^{-6} \times 100 = $ **0.6 mC**
[Check: In a parallel circuit $Q_T = Q_1 + Q_2 + Q_3 + Q_4$
$Q_1 + Q_2 + Q_3 + Q_4 = 0.1 + 0.3 + 0.5 + 0.6 = 1.5$ mC $= Q_T$]

Problem 14 Capacitances of 3 µF, 6 µF and 12 µF are connected in series across a 350 V supply. Calculate (a) the equivalent circuit capacitance, (b) the charge on each capacitor and (c) the pd across each capacitor.

The circuit diagram is shown in *Fig 1*.
(a) The equivalent circuit capacitance C for three capacitors in series is given by:
$$\frac{1}{C} = \frac{1}{C_1} + \frac{1}{C_2} + \frac{1}{C_3}$$
i.e. $\frac{1}{C} = \frac{1}{3} + \frac{1}{6} + \frac{1}{12} = \frac{4+2+1}{12} = \frac{7}{12}$

Hence the equivalent circuit capacitance

$C = \frac{12}{7} = 1\frac{5}{7}\ \mu F$

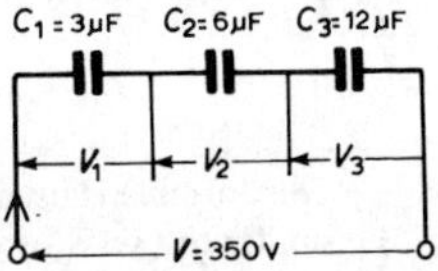

Fig 1

(b) Total charge $Q_T = CV$

Hence $Q_T = \frac{12}{7} \times 10^{-6} \times 350 = 600\ \mu C$ or **0.6 mC**

Since the capacitors are connected in series 0.6 mC is the charge on each of them.

(c) The voltage across the 3 µF capacitor, $V_1 = \frac{Q}{C_1} = \frac{0.6 \times 10^{-3}}{3 \times 10^{-6}} = $ **200 V**

The voltage across the 6 µF capacitor, $V_2 = \frac{Q}{C_2} = \frac{0.6 \times 10^{-3}}{6 \times 10^{-6}} = $ **100 V**

The voltage across the 8 µF capacitor, $V_3 = \frac{Q}{C_3} = \frac{0.6 \times 10^{-3}}{12 \times 10^{-6}} = $ **50 V**

[Check: In a series circuit $V = V_1 + V_2 + V_3$
$V_1 + V_2 + V_3 = 200 + 100 + 50 = 350$ V = supply voltage.]

In practice, capacitors are rarely connected in series unless they are of the same capacitance. The reason for this can be seen from the above problem where the lowest valued capacitor (i.e. 3 μF) has the highest pd across it (i.e. 200 V) which means that if all the capacitors have an identical construction they must all be rated at the highest voltage.

Problem 15 For the arrangement shown in Fig 2 find (a) the equivalent capacitance of the circuit, (b) the voltage across QR, and (c) the charge on each capacitor.

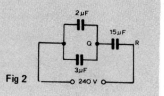

Fig 2

(a) 2 μF in parallel with 3 μF gives an equivalent capacitance of 2 μF + 3 μF = 5 μF. The circuit is now as shown in Fig 3. The equivalent capacitance of 5 μF in series with 15 μF is given by

$$\frac{5 \times 15}{5 + 15} \mu F$$

i.e. $\frac{75}{20}$ or 3.75 μF

Fig 3

(b) The charge on each of the capacitors shown in Fig 3 will be the same since they are connected in series. Let this charge be Q coulombs.
Then $Q = C_1 V_1 = C_2 V_2$
i.e. $\quad\quad 5V_1 = 15V_2$
$\quad\quad\quad V_1 = 3V_2 \quad\quad (1)$
Also $\quad V_1 + V_2 = 240$ V
Hence $\quad 3V_2 + V_2 = 240$ V from (1)
Thus $\quad\quad V_2 = 60$ V and $V_1 = 180$ V

Hence the voltage across QR is 60 V

(c) The charge on the 15 μF capacitor is $C_2 V_2 = 15 \times 10^{-6} \times 60 = $ **0.9 mC**
The charge on the 2 μF capacitor is $2 \times 10^{-6} \times 180 = $ **0.36 mC**
The charge on the 3 μF capacitor is $3 \times 10^{-6} \times 180 = $ **0.54 mC**

Further problems on capacitors connected in parallel and in series may be found in section C(c), Problems 23 to 33, page 45.

(e) PROBLEMS ON ENERGY STORED IN CAPACITORS

Problem 16 (a) Determine the energy stored in a 3 μF capacitor when charged to 400 V.
(b) Find also the average power developed if this energy is dissipated in a time of 10 μs.

(a) Energy stored $W = \frac{1}{2}CV^2$ J

$$= \frac{1}{2} \times 3 \times 10^{-6} \times 400^2 = \frac{3}{2} \times 16 \times 10^{-2} = \textbf{0.24 J}$$

(b) Power $= \dfrac{\text{energy}}{\text{time}} = \dfrac{0.24}{10 \times 10^{-6}}$ W $= \textbf{24 kW}$

Problem 17 A 12 μF capacitor is required to store 4 J of energy. Find the pd to which the capacitor must be charged.

Energy stored $W = \frac{1}{2}CV^2$ hence $V^2 = \dfrac{2W}{C}$

and $V = \sqrt{\left(\dfrac{2W}{C}\right)} = \sqrt{\left(\dfrac{2 \times 4}{12 \times 10^{-6}}\right)} = \sqrt{\left(\dfrac{2 \times 10^6}{3}\right)}$

$$= \textbf{816.5 V}$$

Problem 18 A capacitor is charged with 10 mC. If the energy stored is 1.2 J find (a) the voltage and (b) the capacitance.

Energy stored $W = \frac{1}{2}CV^2$ and $C = \dfrac{Q}{V}$

Hence $W = \frac{1}{2}\left(\dfrac{Q}{V}\right)V^2 = \frac{1}{2}QV$

from which $V = \dfrac{2W}{Q}$

$Q = 10$ mC $= 10 \times 10^{-3}$ C and $W = 1.2$ J

(a) Voltage $V = \dfrac{2W}{Q} = \dfrac{2 \times 1.2}{10 \times 10^{-3}} = \textbf{0.24 kV or 240 V}$

(b) Capacitance $C = \dfrac{Q}{V} = \dfrac{10 \times 10^{-3}}{240}$ F $= \dfrac{10 \times 10^6}{240 \times 10^3}$ μF $= \textbf{41.67 }\mu\textbf{F}$

Problem 19 Compare briefly standard types of capacitor.

1 **Variable air capacitors.** These usually consist of two sets of metal plates (such as aluminium) one fixed, the other variable. The set of moving plates rotate on a

spindle as shown by the end view in *Fig 4*. As the moving plates are rotated through half a revolution, the meshing, and therefore the capacitance, varies from a minimum to a maximum value. Variable air capacitors are used in radio and electronic circuits where very low losses are required, or where a variable capacitance is needed. The maximum value of such capacitors is between 500 pF and 1000 pF.

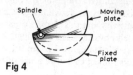

Fig 4

2 **Mica capacitors.** A typical older type construction is shown in *Fig 5*.

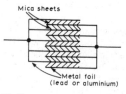

Fig 5

Usually the whole capacitor is impregnated with wax and placed in a bakelite case. Mica is easily obtained in thin sheets and is a good insulator. However, mica is expensive and is not used in capacitors above about 0.1 μF. A modified form of mica capacitor is the silvered mica type. The mica is coated on both sides with a thin layer of silver which forms the plates. Capacitance is stable and less likely to change with age. Such capacitors have a constant capacitance with change of temperature, a high working voltage rating and a long service life and are used in high frequency circuits with fixed values of capacitance up to about 1000 pF.

3 **Paper capacitors.** A typical paper capacitor is shown in *Fig 6* where the length of the roll corresponds to the capacitance required. The whole is usually impregnated with oil or wax to exclude moisture, and then placed in a plastic or aluminium container for protection. Paper capacitors are made in various working voltages up to about 1 μF. Disadvantages of paper capacitors include variation in capacitance with temperature change and a shorter service life than most other types of capacitor.

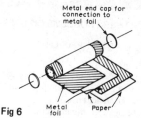

Fig 6

4 **Ceramic capacitors.** These are made in various forms, each type of construction depending on the value of capacitance required. For high values, a tube of ceramic material is used as shown in the cross section of *Fig 7*. For smaller values

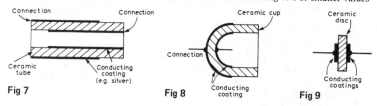

Fig 7 Fig 8 Fig 9

the cup construction is used as shown in *Fig 8*, and for still smaller values the disc construction shown in *Fig 9* is used. Certain ceramic materials have a very high permittivity and this enables capacitors of high capacitance to be made which are of small physical size with a high working voltage rating. Ceramic capacitors are available in the range 1 pF to 0.1 µF and may be used in high frequency electronic circuits subject to a wide range of temperatures.

5 **Plastic capacitors.** Some plastic materials such as polystyrene and Teflon can be used as dielectrics. Construction is similar to the paper capacitor but using a plastic film instead of paper. Plastic capacitors operate well under conditions of high temperature, provide a precise value of capacitance, a very long service life and high reliability.

6 **Electrolytic capacitors.** Construction is similar to the paper capacitor with aluminium foil used for the plates and with a thick absorbent material, such as paper, impregnated with an electrolyte (ammonium borate), separating the plates. The finished capacitor is usually assembled in an aluminium container and hermetically sealed. Its operation depends on the formation of a thin aluminium oxide layer on the positive plate by electrolytic action when a suitable direct potential is maintained between the plates. This oxide layer is very thin and forms the dielectric. (The absorbent paper between the plates is a conductor and does not act as a dielectric.) Such capacitors must always be used on dc and must be connected with the correct polarity; if this is not done the capacitor will be destroyed since the oxide layer will be destroyed. Electrolytic capacitors are manufactured with working voltage from 6 V to 500 V, although accuracy is generally not very high. These capacitors possess a much larger capacitance than other types of capacitors of similar dimensions due to the oxide film being only a few microns thick. The fact that they can be used only on dc supplies limit their usefulness.

C. FURTHER PROBLEMS ON CAPACITORS AND CAPACITANCE

(a) SHORT ANSWER PROBLEMS

1 Explain the term 'electrostatics'.
2 Complete the statements: Like charges . . . . ; unlike charges . . . .
3 How can an 'electric field' be established between two parallel metal plates?
4 What is capacitance?
5 State the unit of capacitance.
6 Complete the statement: Capacitance $= \frac{\ldots}{\ldots}$
7 Complete the statements: (a) 1 µF = . . . F
   (b) 1 pF = . . . F.
8 Complete the statement: Electric field strength $E = \frac{\ldots}{\ldots}$
9 Complete the statement: Electric flux density $D = \frac{\ldots}{\ldots}$
10 Draw the electrical circuit diagram symbol for a capacitor.
11 Name two practical examples where capacitance is present, although undesirable.
12 The insulating material separating the plates of a capacitor is called the . . . .
13 10 volts applied to a capacitor results in a charge of 5 coulombs. What is the capacitance of the capacitor?

14 Three 3 µF capacitors are connected in parallel. The equivalent capacitance is . . . .
15 Three 3 µF capacitors are connected in series. The equivalent capacitance is . . . .
16 State an advantage of series connected capacitors.
17 Name three factors upon which capacitance depends.
18 What does 'relative permittivity' mean?
19 Define 'permittivity of free space'.
20 Name five types of capacitor commonly used.
21 Sketch a typical rolled paper capacitor.
22 Explain briefly the construction of a variable air capacitor.
23 State three advantages and one disadvantage of mica capacitors.
24 Name two disadvantages of paper capacitors.
25 Between what values of capacitance are ceramic capacitors normally available?
26 What main advantages do plastic capacitors possess?
27 Explain briefly the construction of an electrolytic capacitor.
28 What is the main disadvantage of electrolytic capacitors?
29 Name an important advantage of electrolytic capacitors.
30 What safety precautions should be taken when a capacitor is disconnected from a supply?
31 What is meant by the 'dielectric strength' of a material?
32 State the formula used to determine the energy stored by a capacitor.

(b) MULTI-CHOICE PROBLEMS (Answers on page 152)

1 Electrostatics is a branch of electricity concerned with
  (a) energy flowing across a gap between conductors;
  (b) charges at rest;
  (c) charges in motion;
  (d) energy in the form of charges.
2 The capacitance of a capacitor is the ratio
  (a) charge to pd between plates;
  (b) pd between plates to plate spacing;
  (c) pd between plates to thickness of dielectric;
  (d) pd between plates to charge.
3 The pd across a 10 µF capacitor to charge it with 10 mC is
  (a) 100 V; (b) 1 kV; (c) 1 V; (d) 10 V.
4 The charge on a 10 pF capacitor when the voltage applied to it is 10 kV is:
  (a) 100 µC; (b) 0.1 C; (c) 0.1 µC; (d) 0.01 µC.
5 Four 2 µF capacitors are connected in parallel. The equivalent capacitance is
  (a) 8 µF; (b) 0.5 µF.
6 Four 2 µF capacitors are connected in series. The equivalent capacitance is
  (a) 8 µF; (b) 0.5 µF.
7 State which of the following is false.
  The capacitance of a capacitor
  (a) is proportional to the cross sectional area of the plates;
  (b) is proportional to the distance between the plates;
  (c) depends on the number of plates;
  (d) is proportional to the relative permittivity of the dielectric.
8 State which of the following statements is false.
  (a) An air capacitor is normally a variable type.

(b) A paper capacitor generally has a shorter service life than most other types of capacitor.
(c) An electrolytic capacitor must be used only on a.c. supplies.
(d) Plastic capacitors generally operate satisfactorily under conditions of high temperature.

9 The energy stored in a 10 µF capacitor when charged to 500 V is
(a) 1.25 mJ; (b) 0.025 µJ; (c) 1.25 J; (d) 1.25 C.

10 The capacitance of a variable air capacitor is a maximum when
(a) the movable plates half overlap the fixed plates;
(b) the movable plates are most widely separated from the fixed plates;
(c) both sets of plates are exactly meshed;
(d) the movable plates are closer to one side of the fixed plate than to the other.

(c) CONVENTIONAL PROBLEMS

(Where appropriate take ϵ_0 as 8.85×10^{-12} F/m)

Electric flux density and electric field strength

1 A capacitor uses a dielectric 0.04 mm thick and operates at 30 V. What is the electric field strength across the dielectric at this voltage?

[750 kV/m]

2 A two plate capacitor has a charge of 25 C. If the effective area of each plate is 5 cm^2 find the electric flux density of the electric field.

[50 kC/m^2]

3 A charge of 1.5 µC is carried on two parallel rectangular plates each measuring 60 mm by 80 mm. Calculate the electric flux density. If the plates are spaced 10 mm apart and the voltage between them is 0.5 kV determine the electric field strength.

[312.5 µC/m^2; 50 kV/m]

4 Two parallel plates are separated by a dielectric and charged with 10 µC. Given that the area of each plate is 50 cm^2, calculate the electric flux density in the dielectric separating the plates.

[2 mC/m^2]

5 The flux density between two plates separated by polystyrene of relative permittivity 2.5 is 5 µC/m^2. Find the voltage gradient between the plates.

[226 kV/m]

6 Two parallel plates having a pd of 250 V between them are spaced 1 mm apart. Determine the electric field strength. Find also the flux density when the dielectric between the plates is (a) air and (b) mica of relative permittivity 5.

[250 kV/m; (a) 2.213 µC/m^2 (b) 11.063 µC/m^2]

Q = CV problems

7 Find the charge on a 10 µF capacitor when the applied voltage is 250 V.

[2.5 mC]

8 Determine the voltage across a 1000 pF capacitor to charge it with 2 µC.

[2 kV]

9 The charge on the plates of a capacitor is 6 mC when the potential between them is 2.4 kV. Determine the capacitance of the capacitor.

[2.5 µF]

10 For how long must a charging current of 2 A be fed to a 5 µF capacitor to raise the pd between its plates by 500 V.

[1.25 ms]

11 A direct current of 10 A flows into a previously uncharged 5 µF capacitor for 1 ms. Determine the pd between the plates.

[2 kV]

12 A 16 µF capacitor is charged at a constant current of 4 µA for 2 min. Calculate the final pd across the capacitor and the corresponding charge in coulombs.

[30 V; 480 µC]

13 A steady current of 10 A flows into a previously uncharged capacitor for 1.5 ms when the pd between the plates is 2 kV. Find the capacitance of the capacitor.

[7.5 µF]

Parallel plate capacitor

14 A capacitor consists of two parallel plates each of area 0.01 m^2, spaced 0.1 mm in air. Calculate the capacitance in picofarads.

[885 pF]

15 A waxed paper capacitor has two parallel plates, each of effective area 0.2 m^2. If the capacitance is 4000 pF determine the effective thickness of the paper if its relative permittivity is 2.

[0.885 mm]

16 Calculate the capacitance of a parallel plate capacitor having 5 plates, each 30 mm by 20 mm and separated by a dielectric 0.75 mm thick having a relative permittivity of 2.3.

[65.14 pF]

17 How many plates has a parallel plate capacitor having a capacitance of 5 nF, if each plate is 40 mm square and each dielectric is 0.102 mm thick with a relative permittivity of 6.

[7]

18 A parallel plate capacitor is made from 25 plates, each 70 mm by 120 mm interleaved with mica of relative permittivity 5. If the capacitance of the capacitor is 3000 pF determine the thickness of the mica sheet.

[2.97 mm]

19 A capacitor is constructed with parallel plates and has a value of 50 pF. What would be the capacitance of the capacitor if the plate area is doubled and the plate spacing is doubled?

[200 pF]

20 The capacitance of a parallel plate capacitor is 1000 pF. It has 19 plates, each 50 mm by 30 mm separated by a dielectric of thickness 0.40 mm. Determine the relative permittivity of the dielectric.

[1.67]

21 The charge on the square plates of a multiplate capacitor is 80 µC when the potential between them is 5 kV. If the capacitor has twenty-five plates separated by a dielectric of thickness 0.102 mm and relative permittivity 4.8, determine the width of a plate.

[40 mm]

22 A capacitor is to be constructed so that its capacitance is 4250 pF and to operate at a pd of 100 V across its terminals. The dielectric is to be polythene ($\epsilon_r = 2.3$) which, after allowing a safety factor, has a dielectric strength of 20 MV/m. Find (a) the thickness of the polythene needed, and (c) the area of a plate.

[(a) 0.005 mm; (b) 10.44 cm^2]

Capacitors connected in parallel and in series

23 Capacitors of 2 µF and 6 µF are connected (a) in parallel and (b) in series. Determine the equivalent capacitance in each case.

[(a) 8 µF; (b) 1.5 µF]

24 Find the capacitance to be connected in series with a 10 μF capacitor for the equivalent capacitance to be 6 μF.

[15 μF]

25 What value of capacitance would be obtained if capacitors of 0.15 μF and 0.1 μF are connected (a) in series and (b) in parallel.

[(a) 0.06 μF; (b) 0.25 μF]

26 Two 6 μF capacitors are connected in series with one having a capacitance of 12 μF. Find the total equivalent circuit capacitance. What capacitance must be added in series to obtain a capacitance of 1.2 μF?

[2.4 μF; 2.4 μF]

27 Determine the equivalent capacitance when the following capacitors are connected (a) in parallel and (b) in series:
   (i) 2 μF, 4 μF and 8 μF
   (ii) 0.02 μF, 0.05 μF and 0.1 μF
   (iii) 50 pF and 450 pF
   (iv) 0.01 μF and 200 pF

[(a) (i) 14 μF, (ii) 0.17 μF, (iii) 500 pF, (iv) 0.0102 μF
(b) (i) 1$\frac{1}{7}$ μF, (ii) 0.0125 μF, (iii) 45 pF, (iv) 196.1 pF]

28 For the arrangement shown in *Fig 10* find (a) the equivalent circuit capacitance and (b) the voltage across a 4.5 μF capacitor.

[(a) 1.2 μF; (b) 100 V]

29 Three 12 μF capacitors are connected in series across a 750 V supply. Calculate (a) the equivalent capacitance, (b) the charge on each capacitor and (c) the pd across each capacitor.

[(a) 4 μF; (b) 3 mC; (c) 250 V]

30 If two capacitors having capacitances of 3 μF and 5 μF respectively are connected in series across a 240 V supply determine (a) the pd across each capacitor and (b) the charge on each capacitor.

[(a) 150 V, 90 V; (b) 0.45 mC on each]

31 In *Fig 11* capacitors P, Q and R are identical and the total equivalent capacitance of the circuit is 3 μF. Determine the values of P, Q and R.

[4.2 μF each]

32 Capacitances of 4 μF, 8 μF and 16 μF are connected in parallel across a 200 V supply. Determine (a) the equivalent capacitance, (b) the total charge and (c) the charge on each capacitor.

[(a) 28 μF, (b) 5.6 mC, (c) 0.8 mC, 1.6 mC, 3.2 mC]

33 A circuit consists of two capacitors P and Q in parallel, connected in series with another capacitor R. The capacitances of P, Q and R are 4 μF, 12 μF and 8 μF respectively. When the circuit is connected across a 300 V dc supply find (a) the

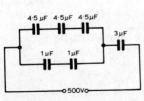

Fig 10

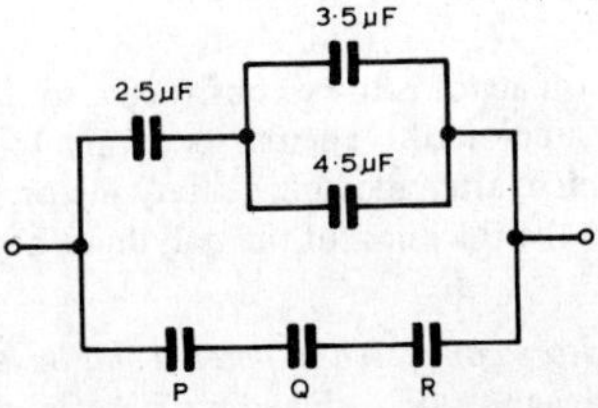

Fig 11

total capacitance of the circuit, (b) the pd across each capacitor and (c) the charge on each capacitor.

$$\begin{bmatrix} \text{(a)} & 5\tfrac{1}{3} \ \mu\text{F, (b) 100 V across P, 100 V across Q, 200 V across R,} \\ \text{(c)} & 0.4 \ \text{mC on P, 1.2 mC on Q, 1.6 mC on R.} \end{bmatrix}$$

Energy stored in capacitors

34 When a capacitor is connected across a 200 V supply the charge is 4 μC. Find (a) the capacitance and (b) the energy stored.

[(a) 0.02 μF; (b) 0.4 mJ]

35 Find the energy stored in a 10 μF capacitor when charged to 2 kV.

[20 J]

36 A 3300 pF capacitor is required to store 0.5 mJ of energy. Find the pd to which the capacitor must be charged.

[550 V]

37 A capacitor is charged with 8 mC. If the energy stored is 0.4 J find (a) the voltage and (b) the capacitance.

[(a) 100 V, (b) 80 μF]

38 A capacitor, consisting of two metal plates each of area 50 cm² and spaced 0.2 mm apart in air, is connected across a 120 V supply. Calculate (a) the energy stored, (b) the electric flux density and (c) the potential gradient.

[(a) 1.593 μJ; (b) 5.31 μC/m²; (c) 600 kV/m]

39 A bakelite capacitor is to be constructed to have a capacitance of 0.04 μF and to have a steady working potential of 1 kV maximum. Allowing a safe value of field stress of 25 MV/m find (a) the thickness of bakelite required, (b) the area of plate required if the relative permittivity of bakelite is 5, (c) the maximum energy stored by the capacitor and (d) the average power developed if this energy is dissipated in a time of 20 μs.

[(a) 0.04 mm; (b) 361.6 cm²; (c) 0.02 J; (d) 1 kW]

4 The magnetic field

A. FORMULAE AND DEFINITIONS ASSOCIATED WITH MAGNETIC CIRCUITS

1 A **permanent magnet** is a piece of ferromagnetic material (such as iron, nickel or cobalt) which has properties of attracting other pieces of these materials.
2 The area around a magnet is called the **magnetic field** and it is in this area that the effects of the **magnetic force** produced by the magnet can be detected.
3 **Magnetic fields** can be established by electric currents as well as by permanent magnets.
4 The **magnetic flux** Φ is the amount of magnetic field (or the number of lines of force) produced by a magnetic source.
5 The **unit of magnetic flux** is the weber, Wb.
6 **Magnetic flux density** B is the amount of flux passing through a defined area that is perpendicular to the direction of the flux.
7 Magnetic flux density $= \dfrac{\text{magnetic flux}}{\text{area}}$, i.e. $B = \Phi/A$
8 The **unit of flux density** is the tesla T, where $1\text{ T} = 1\text{ Wb/m}^2$.
9 **Magnetomotive force (mmf)**

$F_m = NI$ ampere-turns (At)

where N = number of conductors (or turns)

I = current in amperes.

Since 'turns' has no units, the SI unit of mmf is the ampere, but to avoid any possible confusion 'ampere-turns', (At), are used in this chapter.

10 **Magnetic field strength, or magnetising force**

$H = \dfrac{NI}{l}$ At/m where l = mean length of flux path in metres.

Hence, mmf $= NI = Hl$ At

11 For air, or any non-magnetic medium, the ratio of magnetic flux density to magnetising force is a constant, i.e. $B/H =$ a constant. This constant is μ_0, the **permeability of free space** (or the magnetic space constant) and is equal to $4\pi \times 10^{-7}$ H/m.

Hence $\dfrac{B}{H} = \mu_0$

12 For **ferromagnetic mediums**:

$\dfrac{B}{H} = \mu_0 \mu_r$

where μ_r is the relative permeability, and is defined as

$\dfrac{\text{flux density in material}}{\text{flux density in air}}$

Its value varies with the type of magnetic material and since μ_r is a ratio of flux densities, it has no units. From its definition, μ_r for air is 1.

13 $\mu_0 \mu_r = \mu$, called the **absolute permeability**.

14 By plotting measured values of flux density B against magnetic field strength H, a **magnetisation curve** (or **B–H curve**) is produced. For non-magnetic materials this is a straight line. Typical curves for four magnetic materials are shown on page 50.

15 The **relative permeability** of a ferromagnetic material is proportional to the slope of the B–H curve and thus varies with the magnetic field strength. The approximate range of values of relative permeability μ_r for some common magnetic materials are:

Cast iron $\quad \mu_r = 100\text{–}250;$ $\qquad$ Mild steel $\quad \mu_r = 200\text{–}800$
Silicon iron $\quad \mu_r = 1000\text{–}5000;$ $\qquad$ Cast steel $\quad \mu_r = 300\text{–}900$
Mumetal $\quad \mu_r = 200\text{–}5000;$ $\qquad$ Stalloy $\quad \mu_r = 500\text{–}6000$

16 The 'magnetic resistance' of a magnetic circuit to the presence of magnetic flux is called **reluctance**. The symbol for reluctance is S (or R_m).

17 Reluctance, $S = \dfrac{\text{mmf}}{\Phi} = \dfrac{NI}{\Phi} = \dfrac{Hl}{BA} = \dfrac{l}{\dfrac{B}{H}A} = \dfrac{l}{\mu_0 \mu_r A}$

18 The **unit of reluctance** is $1/H$ (or H^{-1}) or At/Wb.

19 For a series magnetic circuit having n parts, the **total reluctance** S is given by:
$S = S_1 + S_2 + \ldots + S_n$.
(This is similar to resistors connected in series in an electrical circuit.)

20 **Comparison between electrical and magnetic quantities.**

Electric circuit	Magnetic circuit
emf E (V)	mmf F_m (A)
current I (A)	flux Φ (Wb)
resistance R (Ω)	reluctance S (H^{-1})
$I = \dfrac{E}{R}$	$\Phi = \dfrac{\text{mmf}}{S}$
$R = \dfrac{\rho l}{A}$	$S = \dfrac{l}{\mu_0 \mu_r A}$

21 **Ferromagnetic materials** have a low reluctance and can be used as magnetic screens to prevent magnetic fields affecting materials within the screen.

22 **Hysteresis** is the 'lagging' effect of flux density B whenever there are changes in the magnetic field strength H.

23 When an initially unmagnetised ferromagnetic material is subjected to a varying magnetic field strength H, the flux density B produced in the material varies as shown in *Fig 1*, the arrows indicating the direction of the cycle. *Fig 1* is known as a **hysteresis loop**.

OX = residual flux density or remanence
OY = coercive force
PP$'$ = saturation flux density.

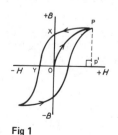

Fig 1

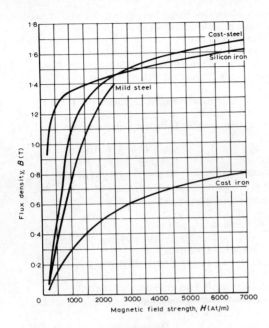

Fig 2 B–H curves for four materials

24 Hysteresis results in a **dissipation of energy** which appears as a heating of the magnetic material. The energy loss associated with hysteresis is proportional to the area of the hysteresis loop.

25 The area of a hysteresis loop varies with the type of material. The area, and thus the energy loss, is much greater for hard materials than for soft materials.

B. WORKED PROBLEMS ON MAGNETIC CIRCUITS

(a) MAGNETIC CIRCUIT QUANTITIES

Problem 1 A magnetic pole face has a rectangular section having dimensions 20 cm by 10 cm. If the total flux emerging from the pole is 150 μWb, calculate the flux density.

Flux $\Phi = 150 \, \mu\text{Wb} = 150 \times 10^{-6}$ Wb
Cross sectional area $A = 20 \times 10 = 200 \text{ cm}^2 = 200 \times 10^{-4} \text{ m}^2$
Flux density $B = \dfrac{\Phi}{A} = \dfrac{150 \times 10^{-6}}{200 \times 10^{-4}} = \textbf{0.0075 T or 7.5 mT}$

Problem 2 A flux density of 1.2 T is produced in a piece of cast steel by a magnetising force of 1250 At/m. Find the relative permeability of the steel under these conditions.

For a magnetic material:

$$B = \mu_0 \mu_r H$$

i.e. $\mu_r = \dfrac{B}{\mu_0 H} = \dfrac{1.2}{(4\pi \times 10^{-7})(1250)} = \mathbf{764}$

Problem 3 Determine the magnetic field strength and the mmf required to produce a flux density of 0.25 T in an air gap of length 12 mm.

For air: $B = \mu_0 H$ (since $\mu_r = 1$)

Magnetic field strength $H = \dfrac{B}{\mu_0} = \dfrac{0.25}{4\pi \times 10^{-7}} = \mathbf{198\,940\ At/m}$

$$\text{mmf} = Hl = 198\,940 \times 12 \times 10^{-3} = \mathbf{2387\ At}$$

Problem 4 A magnetising force of 8000 At/m is applied to a circular magnetic circuit of mean diameter 30 cm by passing a current through a coil wound on the circuit. If the coil is uniformly wound around the circuit and has 750 turns, find the current in the coil.

$H = 8000$ At/m; $l = \pi d = \pi \times 30 \times 10^{-2}$ m; $N = 750$ turns

Since $H = \dfrac{NI}{l}$

then, $I = \dfrac{Hl}{N} = \dfrac{8000 \times \pi \times 30 \times 10^{-2}}{750}$

Thus, current $I = 10.05$ A

Problem 5 A coil of 300 turns is wound uniformly on a ring of non-magnetic material. The ring has a mean circumference of 40 cm and a uniform cross sectional area of 4 cm². If the current in the coil is 5 A, calculate (a) the magnetic field strength, (b) the flux density and (c) the total magnetic flux in the ring.

(a) Magnetic field strength $H = \dfrac{NI}{l} = \dfrac{300 \times 5}{40 \times 10^{-2}} = \mathbf{3750\ At/m}$

(b) For a non-magnetic material $\mu_r = 1$, thus flux density $B = \mu_0 H$
$= 4\pi \times 10^{-7} \times 3750 = \mathbf{4.712\ mT}$

(c) Flux $\Phi = BA = (4.712 \times 10^{-3})(4 \times 10^{-4}) = \mathbf{1.885\ \mu Wb}$

Problem 6 Determine the reluctance of a piece of mumetal of length 150 mm, cross-sectional area 1800 mm² when the relative permeability is 4000. Find also the absolute permeability of the mumetal.

Reluctance $S = \dfrac{l}{\mu_0 \mu_r A} = \dfrac{150 \times 10^{-3}}{(4\pi \times 10^{-7})(4000)(1800 \times 10^{-6})} = $ **16 580/H**

Absolute permeability, $\mu = \mu_0 \mu_r = (4\pi \times 10^{-7})(4000) = $ **5.027 × 10⁻³ H/m**

Problem 7 An iron ring of mean diameter 10 cm is uniformly wound with 2000 turns of wire. When a current of 0.25 A is passed through the coil a flux density of 0.4 T is set up in the iron. Find (a) the magnetising force and (b) the relative permeability of the iron under these conditions.

$l = \pi d = \pi \times 10 \text{ cm} = \pi \times 10 \times 10^{-2} \text{ m}; N = 2000 \text{ turns}; I = 0.25 \text{ A}; B = 0.4 \text{ T}.$

(a) $H = \dfrac{NI}{l} = \dfrac{2000 \times 0.25}{\pi \times 10 \times 10^{-2}} = \dfrac{5000}{\pi} = $ **1592 At/m.**

(b) $B = \mu_0 \mu_r H.$ Hence $\mu_r = \dfrac{B}{\mu_0 H} = \dfrac{0.4}{(4\pi \times 10^{-7})(1592)} = $ **200.**

Problem 8 A mild steel ring has a radius of 50 mm and a cross-sectional area of 400 mm². A current of 0.5 A flows in a coil wound uniformly around the ring and the flux produced is 0.1 mWb. If the relative permeability at this value of current is 200 find (a) the reluctance of the mild steel and (b) the number of turns on the coil.

$l = 2\pi r = 2 \times \pi \times 50 \times 10^{-3} \text{ m}; A = 400 \times 10^{-6} \text{ m}^2; I = 0.5 \text{ A};$
$\Phi = 0.1 \times 10^{-3} \text{ Wb}; \mu_r = 200.$

(a) Reluctance $S = \dfrac{l}{\mu_0 \mu_r A} = \dfrac{2 \times \pi \times 50 \times 10^{-3}}{(4\pi \times 10^{-7})(200)(400 \times 10^{-6})} = $ **3.125 × 10⁶ /H**

(b) $S = \dfrac{\text{mmf}}{\Phi},$

i.e. $\text{mmf} = S\Phi$
$NI = S\Phi$

Hence $N = \dfrac{S\Phi}{I} = \dfrac{3.125 \times 10^6 \times 0.1 \times 10^{-3}}{0.5} = $ **625 turns**

Problem 9 A uniform ring of cast iron has a cross-sectional area of 10 cm² and a mean circumference of 20 cm. Determine the mmf necessary to produce a flux of 0.3 mWb in the ring. The magnetisation curve for cast iron is shown on page 00.

$A = 10 \text{ cm}^2 = 10 \times 10^{-4} \text{ m}^2; l = 20 \text{ cm} = 0.2 \text{ m}; \Phi = 0.3 \times 10^{-3} \text{ Wb}$

Flux density $B = \dfrac{\Phi}{A} = \dfrac{0.3 \times 10^{-3}}{10 \times 10^{-4}} = 0.3 \text{ T}$

From the magnetisation curve for cast iron on page 50, when $B = 0.3$ T, $H = 1000$ At/m.

Hence mmf $= Hl = 1000 \times 0.2 =$ **200 At**

A tabular method could have been used in this problem. Such a solution is shown below.

Part of circuit	Material	Φ Wb	A m²	$B = \dfrac{\Phi}{A}$ T	H from graph	l m	mmf = Hl At
Ring	Cast iron	0.3×10^{-3}	10×10^{-4}	0.3	1000	0.2	200

Problem 10 From the magnetisation curve for cast iron, shown on page 50, derive the curve of μ_r against H.

$$B = \mu_0 \mu_r H, \text{ hence } \mu_r = \frac{B}{\mu_0 H} = \frac{1}{\mu_0} \times \frac{B}{H} = \frac{10^7}{4\pi} \times \frac{B}{H}$$

A number of co-ordinates are selected from the B–H curve and μ_r is calculated for each as shown in the following table.

B (T)	0.04	0.13	0.17	0.30	0.41	0.49	0.60	0.68	0.73	0.76	0.79
H (At/m)	200	400	500	1000	1500	2000	3000	4000	5000	6000	7000
$\mu_r = \dfrac{10^7}{4\pi} \times \dfrac{B}{H}$	159	259	271	239	218	195	159	135	116	101	90

μ_r is plotted against H as shown in *Fig 3*. The curve demonstrates the change that occurs in the relative permeability as the magnetising force increases.

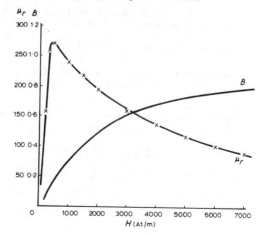

Fig 3 B–H and μ_r–H curves for cast iron

Further problems on magnetic circuit quantities may be found in section C(c), Problems 1 to 18, page 59.

(b) COMPOSITE SERIES MAGNETIC CIRCUITS

Problem 11 A closed magnetic circuit of cast steel contains a 6 cm long path of cross-sectional area 1 cm² and a 2 cm path of cross-sectional area 0.5 cm². A coil of 200 turns is wound around the 6 cm length of the circuit and a current of 0.4 A flows. Determine the flux density in the 2 cm path, if the relative permeability of the cast steel is 750.

For the 6 cm long path:

Reluctance $S_1 = \dfrac{l_1}{\mu_0 \mu_r A_1} = \dfrac{6 \times 10^{-2}}{(4\pi \times 10^{-7})(750)(1 \times 10^{-4})} = 6.366 \times 10^5$ /H

For the 2 cm long path:

Reluctance $S_2 = \dfrac{l_2}{\mu_0 \mu_r A_2} = \dfrac{2 \times 10^{-2}}{(4\pi \times 10^{-7})(750)(0.5 \times 10^{-4})} = 4.244 \times 10^5$ /H

Total circuit reluctance $S = S_1 + S_2 = (6.366 + 4.244) \times 10^5 = 10.61 \times 10^5$ /H

$S = \dfrac{\text{mmf}}{\Phi}$, i.e. $\Phi = \dfrac{\text{mmf}}{S} = \dfrac{NI}{S} = \dfrac{200 \times 0.4}{10.61 \times 10^5} = 7.54 \times 10^{-5}$ Wb

Flux density in the 2 cm path, $B = \dfrac{\Phi}{A} = \dfrac{7.54 \times 10^{-5}}{0.5 \times 10^{-4}} = \mathbf{1.51\ T}$

Problem 12 A silicon iron ring of cross-sectional area 5 cm² has a radial air gap of 2 mm cut into it. If the mean length of the silicon iron path is 40 cm, calculate the magnetomotive force to produce a flux of 0.7 mWb. The magnetisation curve for silicon iron is shown on page 50.

There are two parts to the circuit—the silicon iron and the air gap. The total mmf will be the sum of the mmf's of each part.

For the silicon iron: $B = \dfrac{\Phi}{A} = \dfrac{0.7 \times 10^{-3}}{5 \times 10^{-4}} = 1.4$ T

From the B–H curve for silicon iron on page 50, when $B = 1.4$ T, $H = 1650$ At/m.
Hence the mmf for the iron path $= Hl = 1650 \times 0.4 = 660$ At

For the air-gap:
The flux density will be the same in the air gap as in the iron, i.e. 1.4 T. (This assumes no leakage or fringing occurring.)

For air $H = \dfrac{B}{\mu_0} = \dfrac{1.4}{4\pi \times 10^{-7}} = 1\ 114\ 000$ At/m

Hence the mmf for the air gap $= Hl = 1\ 114\ 000 \times 2 \times 10^{-3} = 2228$ At

Total mmf to produce a flux of 0.6 mWb $= 660 + 2228 = \mathbf{2888\ At}$

A tabular method could have been used as shown below.

Part of circuit	Material	Φ Wb	A m²	B T	H At/m	l m	mmf = Hl At
Ring	Silicon iron	0.7 × 10⁻³	5 × 10⁻⁴	1.4	1650 (from graph)	0.4	660
Air-gap	Air	0.7 × 10⁻³	5 × 10⁻⁴	1.4	$\dfrac{1.4}{4\pi \times 10^{-7}} = 1\,114\,000$	2 × 10⁻³	2228

Total: **2888 At**

Problem 13 *Fig 4* shows a ring formed with two different materials—cast steel and mild steel. The dimensions are:

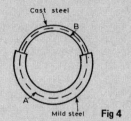

	mean length	cross-sectional area
Mild steel	400 mm	500 mm²
Cast steel	300 mm	312.5 mm²

Find the total mmf required to cause a flux of 500 µWb in the magnetic circuit. Determine also the total circuit reluctance.

Fig 4

A tabular solution is shown below.

Part of circuit	Material	Φ Wb	A m²	B T (= Φ/A)	H At/m (from graphs p.50)	l m	mmf = Hl At
A	Mild steel	500 × 10⁻⁶	500 × 10⁻⁶	1.0	1400	400 × 10⁻³	560
B	Cast steel	500 × 10⁻⁶	312.5 × 10⁻⁶	1.6	4800	300 × 10⁻³	1440

Total: **2000 At**

Total circuit reluctance $S = \dfrac{\text{mmf}}{\Phi} = \dfrac{2000}{500 \times 10^{-6}} = 4 \times 10^6$ /H

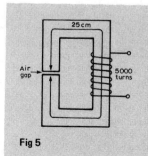

Problem 14 A section through a magnetic circuit of uniform cross-section area 2 cm² is shown in *Fig 5*. The cast steel core has a mean length of 25 cm. The air gap is 1 mm wide and the coil has 5000 turns. The B–H curve for cast steel is shown on page 50. Determine the current in the coil to produce a flux density of 0.80 T in the air gap, assuming that all the flux passes through both parts of the magnetic circuit.

Fig 5

For the cast steel core: When $B = 0.80$ T, $H = 750$ At/m (from page 50)

Reluctance of core $S_1 = \dfrac{l_1}{\mu_0 \mu_r A_1}$ and, since $B = \mu_0 \mu_r H$,

then $\mu_r = \dfrac{B}{\mu_0 H}$. Thus $S_1 = \dfrac{l_1}{\mu_0 \dfrac{B}{\mu_0 H} A} = \dfrac{l_1 H}{BA} = \dfrac{(25 \times 10^{-2})(750)}{(0.8)(2 \times 10^{-4})} = 1\,172\,000/\text{H}$

For the air-gap: Reluctance, $S_2 = \dfrac{l_2}{\mu_0 \mu_r A_2} = \dfrac{l_2}{\mu_0 A_2}$ (since $\mu_r = 1$ for air)

$$= \dfrac{1 \times 10^{-3}}{(4\pi \times 10^{-7})(2 \times 10^{-4})} = 3\,979\,000/\text{H}.$$

Total circuit reluctance $S = S_1 + S_2 = 1\,172\,000 + 3\,979\,000$
$\qquad\qquad\qquad\qquad\qquad\quad = 5\,151\,000$ /H

Flux $\Phi = BA = 0.80 \times 2 \times 10^{-4} = 1.6 \times 10^{-4}$ Wb

$S = \dfrac{\text{mmf}}{\Phi}$

Thus mmf $= S\Phi$
Hence $NI = S\Phi$

Hence current $I = \dfrac{S\Phi}{N} = \dfrac{(5\,151\,000)(1.6 \times 10^{-4})}{5000} = \mathbf{0.165\ A}$

Further problems on composite series magnetic circuits may be found in Section C(c), Problems 19 to 24, page 59.

C. FURTHER PROBLEMS ON MAGNETIC CIRCUITS

(a) SHORT ANSWER PROBLEMS

1 Define magnetic flux.
2 The symbol for magnetic flux is .... and the unit of flux is the ....
3 Define magnetic flux density.
4 The symbol for magnetic flux density is .... and the unit of flux density is ....
5 The symbol for mmf is .... and the unit of mmf is the ....
6 Another name for the magnetising force is .... ; its symbol is .... and its unit is ....
7 Complete the statement: For magnetic materials $\dfrac{\text{flux density}}{\text{magnetic field strength}} = \ldots\ldots$
8 What is absolute permeability?
9 The value of the permeability of free space is ....
10 What is a magnetisation curve?
11 The symbol for reluctance is .... and the unit of reluctance is ....
12 Make a comparison between magnetic and electrical quantities.
13 What is hysteresis?
14 Draw a typical hysteresis loop and identify (a) saturation flux density, (b) remanence and (c) coercive force.

15 State the units of (a) remanence, (b) coercive force.
16 How is magnetic screening achieved?
17 Complete the statement: Magnetic materials have a .... reluctance; non-magnetic materials have a .... reluctance.
18 What loss is associated with hysteresis?

(b) MULTI-CHOICE PROBLEMS (Answers on page 152)

1 The unit of magnetic flux density is the:
(a) weber; (b) weber per metre; (c) ampere per metre; (d) tesla.
2 The total flux in the core of an electrical machine is 20 mWb and its flux density is 1 T. The cross-sectional area of the core is (a) 0.05 m^2; (b) 0.02 m^2; (c) 20 m^2; (d) 50 m^2.

A coil of 100 turns is wound uniformly on a wooden ring. The ring has a mean circumference of 1 m and a uniform cross-sectional area of 10 cm^2. The current in the coil is 1 A.

In *Problems 3 to 7* select the correct answer for each of the required quantities from the following list.

(a) 40π mT; (b) 100 A; (c) $4\pi \times 10^{-10}$/H; (d) 100 At; (e) 0.01 At/m; (f) 40π μT; (g) 40π mWb; (h) $\frac{2.5}{\pi} \times 10^9$ /H; (i) 100 At/m; (j) 0.04π μWb.

3 Magnetomotive force.
4 Magnetic field strength.
5 Magnetic flux density.
6 Magnetic flux.
7 Reluctance.
8. Which of the following statements is false?
(a) For non-magnetic materials reluctance is high.
(b) Energy loss due to hysteresis is greater for harder magnetic materials than for softer magnetic materials.
(c) The remanence of a ferrous material is measured in ampere-turns/metre.
(d) Absolute permeability is measured in henrys per metre.
9 The current flowing in a 500 turn coil wound on an iron ring is 4 A. The reluctance of the circuit is 2×10^6 /H. The flux produced is (a) 1 Wb; (b) 1000 Wb; (c) 1 mWb; (d) 62.5 μWb.
10 A comparison can be made between magnetic and electrical quantities. From the following list, match the magnetic quantities with their equivalent electrical quantities.
(a) Current; (b) reluctance; (c) emf; (d) flux; (e) mmf; (f) resistance.

(c) CONVENTIONAL PROBLEMS

(Where appropriate, assume $\mu_0 = 4\pi \times 10^{-7}$ H/m)

Magnetic circuit quantities

1 What is the flux density in a magnetic field of cross-sectional area 20 cm^2 having a flux of 3 mWb?

[1.5 T]

2. (a) Determine the flux density produced in an air-cored solenoid due to a uniform magnetic field strength of 8000 At/m.
   (b) Iron having a relative permeability 150 at 8000 At/m is inserted into the solenoid of part (a). Find the flux density now in the solenoid.
   [(a) 10.05 mT; (b) 1.508 T]

3. Find the relative permeability of a material if the absolute permeability is 4.084×10^{-4} H/m.
   [325]

4. Find the relative permeability of a piece of silicon iron if a flux density of 1.3 T is produced by a magnetic field strength of 700 At/m.
   [1478]

5. Determine the total flux emerging from a magnetic pole face having dimensions 5 cm by 6 cm, if the flux density is 0.9 T.
   [2.7 mWb]

6. The maximum working flux density of a lifting electromagnet is 1.8 T and the effective area of a pole face is circular in cross-section. If the total magnetic flux produced is 353 mWb determine the radius of the pole face.
   [25 cm]

7. A solenoid 20 cm long is wound with 500 turns of wire. Find the current required to establish a magnetising force of 2500 At/m inside the solenoid.
   [1 A]

8. An electromagnet of square cross-section produces a flux density of 0.45 T. If the magnetic flux is 720 μWb find the dimensions of the electromagnet cross-section.
   [4 cm by 4 cm]

9. Find the magnetic field strength and the magnetomotive force needed to produce a flux density of 0.33 T in an air-gap of length 15 mm.
   [(a) 262 600 At/m; (b) 3939 At]

10. An air-gap between two pole pieces is 20 mm in length and the area of the flux path across the gap is 5 cm^2. If the flux required in the air-gap is 0.75 mWb find the mmf necessary.
    [23 870 At]

11. Find the magnetic field strength applied to a magnetic circuit of mean length 50 cm when a coil of 400 turns is applied to it carrying a current of 1.2 A.
    [960 At/m]

12. A magnetic field strength of 5000 At/m is applied to a circular magnetic circuit of mean diameter 250 mm. If the coil has 500 turns find the current in the coil.
    [7.85 A]

13. Part of a magnetic circuit is made from steel of length 120 mm, cross-sectional area 15 cm^2 and relative permeability 800. Calculate (a) the reluctance and (b) the absolute permeability of the steel.
    [(a) 79 580 /H; (b) 1 mH/m]

14. A steel ring of mean diameter 120 mm is uniformly wound with 1500 turns of wire. When a current of 0.30 A is passed through the coil a flux density of 1.5 T is set up in the steel. Find the relative permeability of the steel under these conditions.
    [1000]

15 A mild steel closed magnetic circuit has a mean length of 75 mm and a cross-sectional area of 320.2 mm². A current of 0.40 A flows in a coil wound uniformly around the circuit and the flux produced is 200 μWb. If the relative permeability of the steel at this value of current is 400 find (a) the reluctance of the material and (b) the number of turns of the coil.

[(a) 466 000 /H; (b) 233]

16 A uniform ring of cast steel has a cross-sectional area of 5 cm² and a mean circumference of 15 cm. Find the current required in a coil of 1200 turns wound on the ring to produce a flux of 0.8 mWb. (Use the magnetisation curve for cast steel shown on page 50.)

[0.60 A]

17 (a) A uniform mild steel ring has a diameter of 50 mm and a cross-sectional area of 1 cm². Determine the mmf necessary to produce a flux of 50 μWb in the ring. Use the B–H curve for mild steel shown on page 50.
(b) If a coil of 440 turns is wound uniformly around the ring in part (a) what current would be required to produce the flux?

[(a) 110 At; (b) 0.25 A]

18 From the magnetisation curve for mild steel shown on page 50, derive the curve of relative permeability against magnetic field strength. From your graph determine (a) the value of μ_r when the magnetic field strength is 1200 At/m, and (b) the value of the magnetic field strength when μ_r is 500.

[(a) 590–600; (b) 2000]

Composite series magnetic circuits

19 A magnetic circuit of cross-sectional area 0.4 cm² consists of one part 3 cm long of material having relative permeability 1200, and a second part 2 cm long of material having relative permeability 750. With a 100 turn coil carrying 2 A, find the value of flux existing in the circuit.

[0.195 mWb]

20 (a) A cast steel ring has a cross-sectional area of 600 mm² and a radius of 25 mm. Determine the mmf necessary to establish a flux of 0.8 mWb in the ring. Use the B–H curve for cast steel shown on page 50.
(b) If a radial air gap 1.5 mm wide is cut in the ring of part (a) find the mmf now necessary to maintain the same flux in the ring.

[(a) 267 At; (b) 1859 At]

21 A closed magnetic circuit made of silicon iron consists of a 40 mm long path of cross-sectional area 90 mm² and a 15 mm long path of cross-sectional area 70 mm². A coil of 50 turns is wound around the 40 mm length of the circuit and a current of 0.39 A flows. Find the flux density in the 15 mm length path if the relative permeability of the silicon iron at this value of magnetising force is 3000.

[1.59 T]

22 For the magnetic circuit shown in *Fig 6* find the current I in the coil needed to produce a flux of 0.45 mWb in the air-gap. The silicon iron magnetic circuit has a uniform cross-sectional area of 3 cm² and its magnetisation curve as shown on page 50.

[0.83 A]

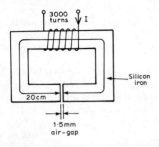

Fig 6

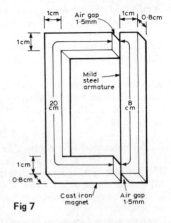

Fig 7

23 A ring forming a magnetic circuit is made from two materials; one part is mild steel of mean length 25 cm and cross-sectional area 4 cm^2, and the remainder is cast iron of mean length 20 cm and cross-sectional area 7.5 cm^2. Use a tabular approach to determine the total mmf required to cause a flux of 0.30 mWb in the magnetic circuit. Find also the total reluctance of the circuit. Use the magnetisation curves shown on page 50.

[540 At; 18×10^5 /H]

24 *Fig 7* shows the magnetic circuit of a relay. When each of the air gaps are 1.5 mm wide find the mmf required to produce a flux density of 0.75 T in the air gaps. Use the *B–H* curves shown on page 50.

[2990 At]

5 Electromagnetic induction

A. FORMULAE AND DEFINITIONS ASSOCIATED WITH ELECTROMAGNETIC INDUCTION

1. When a conductor is moved across a magnetic field, an electromotive force (emf) is produced in the conductor. If the conductor forms part of a closed circuit then the emf produced causes an electric current to flow round the circuit. Hence an emf (and thus current), is 'induced' in the conductor as a result of its movement across the magnetic field. This effect is known as '**electromagnetic induction**'.

2. **Faraday's laws** of electromagnetic induction state:
   (i) 'An induced emf is set up whenever the magnetic field linking that circuit changes.'
   (ii) 'The magnitude of the induced emf in any circuit is proportional to the rate of change of the magnetic flux linking the circuit.'

3. **Lenz's law** states:
   'The direction of an induced emf is always such as to oppsote the effect producing it.'

4. An alternative method to Lenz's law of determining relative directions is given by Fleming's *R*ight-hand rule (often called the gene*R*ator rule) which states:
   'Let the thumb, first finger and second finger of the right hand be extended such that they are all at right angles to each other, as shown in *Fig 1*. If the first finger points in the direction of the magnetic field, the thumb points in the direction of motion of the conductor relative to the magnetic field, then the second finger will point in the direction of the induced emf.'
   Summarising:

   *F*irst finger
   — *F*ield
   Thu*M*b
   — *M*otion
   S*E*cond finger
   — *E*mf

   Fig 1

5. In a **generator**, conductors forming an electric circuit are made to move through a magnetic field. By Faraday's law an emf is induced in the conductors and thus a source of emf is created. A generator converts mechanical energy into electrical energy.

6 The **induced emf** E set up between the ends of the conductor shown in *Fig 2* is given by: $E = Blv$ **volts**,

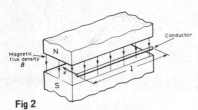

Fig 2

where B, the flux density, is measured in teslas, l, the length of conductor in the magnetic field, is measured in metres, and v, the conductor velocity, is measured in metres per second. If the conductor moves at an angle $\theta°$ to the magnetic field (instead of at 90° as assumed above) then
$E = Blv \sin \theta$.

7 If B teslas is the magnetic flux density, I amperes the current in the conductor and l metres the length of conductor in the magnetic field, then the force F on the current carrying conductor lying at right angles to the direction of the magnetic field is given by:

$F = BIl$ **newtons**

If the conductor and field are at an angle $\theta°$ to each other then:

$F = BIl \sin \theta$ **newtons**

8 The flow of current in a conductor results in a magnetic field around the conductor, the direction of the magnetic field being given by the **screw rule**, which states:

'If a normal right-hand thread screw is screwed along the conductor in the direction of the current, the direction of rotation of the screw is in the direction of the magnetic field.'

This rule is illustrated in *Fig 3*.

9 If the current-carrying conductor shown in *Fig 3(a)* is placed in the magnetic field shown in *Fig 4(a)* then the two fields interact and cause a force to be exerted on the conductor as shown in *Fig 4(b)*. The field is strengthened above the conductor and weakened below, thus tending to move the conductor downwards. This is the basic principle of operation of the **electric motor**. The forces experienced by a number of such conductors can produce motion. A motor is a device that takes in electrical energy and transfers it into mechanical energy.

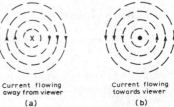

Current flowing away from viewer
(a)

Current flowing towards viewer
(b)

Fig 3 (a) current flowing away from viewer
(b) current flowing towards viewer

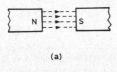

(a)

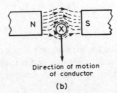

Direction of motion of conductor
(b)

Fig 4

10 The direction of the force exerted on a conductor can be predetermined by using **Fleming's left-hand rule** (often called the **motor rule**), which states:

'Let the thumb, first finger and second finger of the left-hand be extended such that they are all at right angles to each other, as shown in *Fig 5*. If the first finger points in the direction of the magnetic field, the second finger points in the direction of the current, then the thumb will point in the direction of the motion of the conductor.'

Summarising:
*F*irst finger — Field
Se*C*ond finger — Current
Thu*M*b — Motion

Fig 5

11 **Inductance** is the name given to the property of a circuit whereby there is an emf induced into the circuit by the change of flux linkages produced by a current change.
   (i) When the emf is induced in the same circuit as that in which the current is changing, the property is called **self inductance**, L.
   (ii) When the emf is induced in a circuit by a change of flux due to current changing in an adjacent circuit, the property is called **mutual inductance**, M.

12 The **unit of inductance** is the henry, H.
   'A circuit has an inductance of one henry when an emf of one volt is induced in it by a current changing at the rate of one ampere per second.'

13 (i) Induced emf in a coil of N turns, $E = N(\Delta\Phi/t)$ **volts**, where $\Delta\Phi$ is the change in flux, in Webers, and t is the time taken for the flux to change, in seconds.
   (ii) Induced emf in a coil of inductance L henrys, $E = L(\Delta I/t)$ **volts**, where ΔI is the change in current, in amperes, and t is the time taken for the current to change, in seconds.

14 If a current changing from 0 to I amperes, produces a flux change from 0 to Φ Webers then $\Delta I = I$ and $\Delta\Phi = \Phi$. Then, from para 13, induced emf

$$E = \frac{N\Phi}{t} = \frac{LI}{t}$$, from which, **inductance of coil**, $L = N\Phi/I$ henrys.

15 From chapter 4, para. 17 (see page 49) reluctance $S = \frac{\text{mmf}}{\text{flux}} = \frac{NI}{\Phi}$

from which $\Phi = \frac{NI}{S}$

Hence, inductance of coil $L = \frac{N\Phi}{I} = \frac{N}{I}\left(\frac{NI}{S}\right) = \frac{N^2}{S}$, i.e. $L \propto N^2$

16 **Mutually induced emf** in the second coil
$$E_2 = M\frac{\Delta I_1}{t}, \text{ volts}$$

where M is the mutual inductance between two coils, in henrys,
ΔI_1 is the change in current in the first coil, in amperes
t is the time the current takes to change in the first coil, in seconds.

17 **A transformer** is a device which uses the phenomenon of mutual inductance to change the value of alternating voltages. A transformer is represented in *Fig 6(a)* and its circuit diagram symbol is shown in *Fig 6(b)*. When an alternating voltage is applied to the primary winding, an alternating current is produced in the winding. This current produces an alternating flux in the core which links with the secondary winding. The flux induces an alternating emf in the secondary winding by mutual induction.

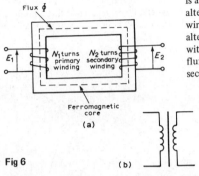

Fig 6

Since $E = \dfrac{N\Phi}{t}$

then $E_1 = \dfrac{N_1 \Phi}{t}$ and $E_2 = \dfrac{N_2 \Phi}{t}$

For an ideal transformer (i.e. no losses),

$\dfrac{E_1}{N_1} = \dfrac{E_2}{N_2}$ i.e. $\dfrac{E_1}{E_2} = \dfrac{N_1}{N_2}$

$\dfrac{E_1}{E_2}$ is called the voltage ratio and $\dfrac{N_1}{N_2}$ the turns ratio.

If $N_2 < N_1$ then $E_2 < E_1$ and the device is termed a step down transformer.
If $N_2 > N_1$ then $E_2 > E_1$ and the device is termed a step up transformer.

18 The energy W stored in the magnetic field of an inductor is given by:

$W = \dfrac{1}{2} LI^2$ joules

B. WORKED PROBLEMS ON ELECTROMAGNETIC INDUCTION

(a) DETERMINATION OF THE FORCE AND DIRECTION ON A CURRENT-CARRYING CONDUCTOR IN A MAGNETIC FIELD

Problem 1 A conductor carries a current of 20 A and is at right angles to a magnetic field having a flux density of 0.9 T. If the length of the conductor in the field is 30 cm, calculate the force acting on the conductor. Determine also the value of the force if the conductor is inclined at an angle of $30°$ to the direction of the field.

$B = 0.9$ T; $I = 20$ A; $l = 30$ cm $= 0.30$ m
Force $F = BIl = (0.9)(20)(0.30)$ newtons when the conductor is at right angles to the field (as shown in *Fig 7(a)*), i.e. $F = 5.4$ N
When the conductor is inclined at $30°$ to the field (as shown in *Fig 7(b)*) then
Force $F = BIl \sin \theta$
i.e. $F = (0.9)(20)(0.30) \sin 30°$ $F = 2.7$ N

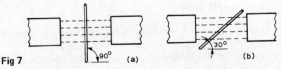

Fig 7

Problem 2 Determine the current required in a 400 mm length of conductor of an electric motor, when the conductor is situated at right angles to a magnetic field of flux density 1.2 T, if a force of 1.92 N is to be exerted on the conductor. If the conductor is vertical, the current flowing downwards and the direction of the magnetic field is from left to right, what is the direction of the force?

$F = 1.92$ N; $l = 400$ mm $= 0.40$ m; $B = 1.2$ T
Since
$F = BIl$, $\qquad I = \dfrac{F}{Bl}$

Hence current $\quad I = \dfrac{1.92}{(1.2)(0.4)} = 4$ A

If the current flows downwards, the direction of its magnetic field due to the current alone will be clockwise when viewed from above. The lines of flux will reinforce (i.e. strengthen) the main magnetic field at the back of the conductor and will be in opposition in the front (i.e. weaken the field). *Hence the force on the conductor will be from back to front (i.e. towards the viewer).* This direction may also have been deduced using Fleming's left-hand rule.

Problem 3 A conductor 350 mm long carries a current of 10 A and is at right angles to a magnetic field lying between two circular pole faces each of radii 60 mm. If the total flux between the pole faces is 0.5 mWb, calculate the force exerted on the conductor.

$l = 350$ mm $= 0.35$ m; $I = 10$ A; Area of pole face $A = \pi r^2 = \pi(0.06)^2$ m^2
$\Phi = 0.5$ mWb $= 0.5 \times 10^{-3}$ Wb

Force $F = BIl$ and $B = \dfrac{\Phi}{A}$

Hence $F = \dfrac{\Phi}{A} Il = \dfrac{(0.5 \times 10^{-3})}{\pi(0.06)^2} (10)(0.35)$ N

i.e. **force = 0.155 N**

Problem 4 With reference to *Fig 8* determine (a) the direction of the force on the conductor in *Fig 8(a)*, (b) the direction of the force on the conductor in *Fig 8(b)*, (c) the direction of the current in *Fig 8(c)* and (d) the polarity of the magnetic system in *Fig 8(d)*.

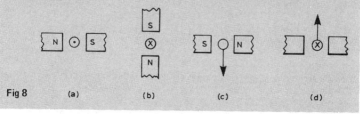

Fig 8   (a)   (b)   (c)   (d)

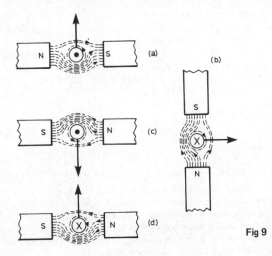

Fig 9

(a) The direction of the main magnetic field is from north to south, i.e. left to right. The current is flowing towards the viewer, and using the screw rule, the direction of the field is anticlockwise. Hence either by Fleming's left-hand rule, or by sketching the interacting magnetic field as shown in *Fig 9(a)*, the direction of the force on the conductor is seen to be upwards.
(b) Using a similar method to part (a) it is seen that the force on the conductor is to the right (see *Fig 9(b)*).
(c) Using Fleming's left-hand rule, or by sketching as in *Fig 9(c)*, it is seen that the current is towards the viewer, i.e. out of the paper.
(d) Similar to part (c), the polarity of the magnetic system is as shown in *Fig 9(d)*.

Further problems on determining the force and direction on a current carrying conductor in a magnetic field may be found in section C(c), Problems 1 to 5, page 75.

(b) MAGNITUDE AND DIRECTIONS OF INDUCED EMF'S

Problem 5 Determine the emf induced in a coil of 200 turns when there is a change of flux of 25 mWb linking with it in 50 ms.

Induced emf $E = N \left(\dfrac{\Delta \Phi}{t}\right) = (200) \left(\dfrac{25 \times 10^{-3}}{50 \times 10^{-3}}\right) = \mathbf{100\ V}$

Problem 6 A flux of 400 µWb passing through a 150 turn coil is reversed in 40 ms. Find the average emf induced.

Since the flux reverses, the flux changes from +400 µWb to −400 µWb, i.e. a change of flux of 800 µWb.

Induced emf $E = N \left(\dfrac{\Delta\Phi}{t}\right) = (150) \left(\dfrac{800 \times 10^{-6}}{40 \times 10^{-3}}\right) = \dfrac{800 \times 150 \times 10^3}{40 \times 10^6}$

Hence, the average emf induced $E = 3$ V

Problem 7 Calculate the emf induced in a coil of inductance 12 H by a current changing at the rate of 4 A/s.

Induced emf $E = L \left(\dfrac{\Delta I}{t}\right) = (12)(4) =$ **48 V**

Problem 8 An emf of 1.5 kV is induced in a coil when a current of 4 A collapses uniformly to zero in 8 ms. Determine the inductance of the coil.

Change in current $\Delta I = 4-0 = 4$ A; $t = 8$ ms $= 8 \times 10^{-3}$ s;

$\dfrac{\Delta I}{t} = \dfrac{4}{8 \times 10^{-3}} = \dfrac{4000}{8} = 500$ A/s; $E = 1.5$ kV $= 1500$ V

Since $E = L \left(\dfrac{\Delta I}{t}\right)$ then $L = \dfrac{E}{\left(\dfrac{\Delta I}{t}\right)} = \dfrac{1500}{500} = 3$ H

Problem 9 An average emf of 40 V is induced in a coil of inductance 150 mH when a current of 6 A is reversed. Calculate the time taken for the current to reverse.

$E = 40$ V; $L = 150$ mH $= 0.15$ H;
Change in current, $\Delta I = 6-(-6) = 12$ A (since the current is reversed)
Since $E = L \left(\dfrac{\Delta I}{t}\right)$ then time $t = \dfrac{L\Delta I}{E} = \dfrac{(0.15)(12)}{40} =$ **0.045 s or 45 ms**

Problem 10 A conductor 300 mm long moves at a uniform speed of 4 m/s at right angles to a uniform magnetic field of flux density 1.25 T. Determine the current flowing in the conductor when (a) its ends are open-circuited, and (b) its ends are connected to a load of 20 Ω resistance.

When a conductor moves in a magnetic field it will have an emf induced in it but this emf can only produce a current if there is a closed circuit.

Induced emf $E = Blv = (1.25) \left(\dfrac{300}{1000}\right) (4) = 1.5$ V

(a) If the ends of the conductor are open circuited **no current will flow** even though 1.5 V has been induced.

(b) From Ohm's law $I = \dfrac{E}{R} = \dfrac{1.5}{20} = 0.075$ A or **75 mA**.

Problem 11 At what velocity must a conductor 75 mm long cut a magnetic field of flux density 0.6 T if an emf of 9 V is to be induced in it? Assume the conductor, the field and the direction of motion are mutually perpendicular.

Induced emf $E = Blv$

Hence velocity $v = \dfrac{E}{Bl}$

Hence $v = \dfrac{9}{(0.6)(75 \times 10^{-3})} = \dfrac{9 \times 10^3}{0.6 \times 75} = \mathbf{200\ m/s}$

Problem 12 A conductor moves with a velocity of 15 m/s at an angle of (a) 90°, (b) 60° and (c) 30° to a magnetic field produced between two square-faced poles of side length 2 cm. If the flux leaving a pole face is 5 μWb find the magnitude of the induced emf in each case.

$V = 15$ m/s; length of conductor in magnetic field, $l = 2$ cm $= 0.02$ m;
$A = 2 \times 2$ cm$^2 = 4 \times 10^{-4}$ m^2; $\Phi = 5 \times 10^{-6}$ Wb.

(a) $E_{90} = Blv \sin 90° = \dfrac{\Phi}{A} lv = \dfrac{(5 \times 10^{-6})}{(4 \times 10^{-4})}(0.02)(15) = \mathbf{3.75\ mV}$

(b) $E_{60} = Blv \sin 60° = E_{90} \sin 60° = 3.75 \sin 60° = \mathbf{3.25\ mV}$

(c) $E_{30} = Blv \sin 30° = E_{90} \sin 30° = 3.75 \sin 30° = \mathbf{1.875\ mV}$

Problem 13 The wing span of a metal aeroplane is 36 m. If the aeroplane is flying at 400 km/h, determine the emf induced between the wing tips. Assume the vertical component of the earth's magnetic field is 40 μT.

Induced emf across wing tips $E = Blv$

$B = 40$ μT $= 40 \times 10^{-6}$ T; $l = 36$ m

$v = 400\ \dfrac{\text{km}}{\text{h}} \times 1000\ \dfrac{\text{m}}{\text{km}} \times \dfrac{1\ \text{h}}{60 \times 60\ \text{s}} = \dfrac{(400)(1000)}{3600} = \dfrac{4000}{36}$ m/s

Hence $E = (40 \times 10^{-6})(36)\dfrac{4000}{36} = \mathbf{0.16\ V}$

Problem 14 The diagram shown in *Fig 10* represents the generation of emf's. Determine
(i) the direction in which the conductor has to be moved in *Fig 10(a)*,
(ii) the direction of the induced emf in *Fig 10(b)*, and
(iii) the polarity of the magnetic system in *Fig 10(c)*.

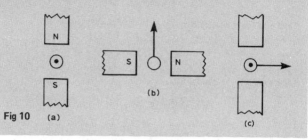

Fig 10 (a) (b) (c)

The direction of the emf, and thus the current due to the emf, may be obtained either by Lenz's law or by Fleming's *R*ight-hand rule (i.e. Gene*R*ator rule).

(i) *Using Lenz's law*: The field due to the magnetic and the field due to the current carrying conductor are shown in *Fig 11(a)* and are seen to reinforce to the left of the conductor. Hence the force on the conductor is to the right. However Lenz's law says that the direction of the induced emf is always such as to oppose the effect producing it. *Thus the conductor will have to be moved to the left.*

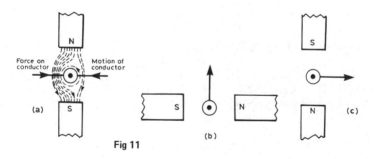

Fig 11

(ii) *Using Fleming's right-hand rule*:
First finger   — *F*ield, i.e. N → S, i.e. right to left
Thu*M*b   — *M*otion, i.e. upwards
S*E*cond finger — *E*mf, i.e. **towards the viewer or out of the paper**, as shown in *Fig 11(b)*.

(iii) The polarity of the magnetic system of *Fig 10(c)* is shown in *Fig 11(c)* and is obtained using Fleming's right-hand rule.

Further problems on the magnitude and direction of induced emf's may be found in section C(c), Problems 6 to 19, page 75.

(c) INDUCTANCE

Problem 15 Calculate the coil inductance when a current of 4 A in a coil of 800 turns produces a flux of 5 mWb linking with the coil.

For a coil, inductance $L = \dfrac{N\Phi}{I} = \dfrac{(800)(5 \times 10^{-3})}{4} = 1\text{ H}$

Problem 16 When a current of 1.5 A flows in a coil the flux linking with the coil is 90 μWb. If the coil inductance is 0.60 H calculate the number of turns of the coil.

For a coil, $L = \dfrac{N\Phi}{I}$. Thus $N = \dfrac{LI}{\Phi} = \dfrac{(0.60)(1.5)}{90 \times 10^{-6}} = 10\,000\text{ turns}$

Problem 17 When carrying a current of 3 A, a coil of 750 turns has a flux of 12 mWb linking with it. Calculate the coil inductance and the emf induced in the coil when the current collapses to zero in 18 ms.

Coil inductance $L = \dfrac{N\Phi}{I} = \dfrac{(750)(12 \times 10^{-3})}{3} = 3\text{ H}$

Induced emf $E = L\left(\dfrac{\Delta I}{t}\right) = 3\left(\dfrac{3-0}{18 \times 10^{-3}}\right) = 500\text{ V}$

(Alternatively $E = N\left(\dfrac{\Delta\Phi}{t}\right) = (750)\left(\dfrac{12 \times 10^{-3}}{18 \times 10^{-3}}\right) = 500\text{ V}$)

Problem 18 A coil has 200 turns and an inductance of 2 mH. How many turns would be needed to produce a 1.28 mH coil (assuming that the same core is used).

Inductance $\quad L \propto N^2$

Hence $\quad \dfrac{L_1}{L_2} = \dfrac{N_1^2}{N_2^2}$

i.e. $\quad \dfrac{2 \times 10^{-3}}{1.28 \times 10^{-3}} = \dfrac{(200)^2}{N_2^2}$

$$N_2 = \sqrt{\left[(200)^2 \left(\dfrac{1.28}{2}\right)\right]} = \mathbf{160\text{ turns}}$$

Further problems on inductance may be found in section C(c), Problems 20 to 25, page 76.

(d) MUTUAL INDUCTANCE

Problem 19 Calculate the mutual inductance between two coils when a current changing at 200 A/s in one coil induces an emf of 1.5 V in the other.

Induced emf $E_2 = M\left(\dfrac{\Delta I_1}{t}\right)$ Hence $1.5 = M(200)$

Thus mutual inductance, $M = \dfrac{1.5}{200} = 0.0075$ H or 7.5 mH

Problem 20 The mutual inductance between two coils is 18 mH. Calculate the steady rate of change of current in one coil to induce an emf of 0.72 V in the other.

Induced emf $E_2 = M\left(\dfrac{\Delta I_1}{t}\right)$

Hence rate of change of current $\dfrac{\Delta I_1}{t} = \dfrac{E_2}{M} = \dfrac{0.72}{0.018} = 40$ A/s

Problem 21 Two coils have a mutual inductance of 0.2 H. If the current in one coil is changed from 10 A to 4 A in 10 ms calculate (a) the average induced emf in the second coil, and (b) the change of flux linked with the second coil if it is wound with 500 turns.

(a) Induced emf $E_2 = M\left(\dfrac{\Delta I_1}{t}\right) = (0.2)\left(\dfrac{10-4}{10 \times 10^{-3}}\right) = 120$ V

(b) Induced emf $E = N\left(\dfrac{\Delta \Phi}{t}\right)$, hence $\Delta \Phi = \dfrac{Et}{N}$

Thus the change of flux $\Delta \Phi = \dfrac{120(10 \times 10^{-3})}{500} = 2.4$ mWb

Further problems on mutual inductance may be found in section C(c), Problems 26 to 30, page 77.

(e) THE TRANSFORMER

Problem 22 A transformer has 500 primary turns and 3000 secondary turns. If the primary voltage is 240 V determine the secondary voltage, assuming an ideal transformer.

For an ideal transformer, voltage ratio = turns ratio

i.e. $\dfrac{E_1}{E_2} = \dfrac{N_1}{N_2}$ Hence $\dfrac{240}{E_2} = \dfrac{500}{3000}$

Hence secondary voltage $E_2 = \dfrac{(3000)(240)}{(500)} = 1440$ V or 1.44 kV

Problem 23 An ideal transformer with a turns ratio of 2:7 is fed from a 240 V supply. Determine its output voltage.

A turns ratio of 2:7 means that the transformer has 2 turns on the primary for every 7 turns on the secondary (i.e. a step-up transformer).

Thus $\dfrac{N_1}{N_2} = \dfrac{2}{7}$

For an ideal transformer $\dfrac{N_1}{N_2} = \dfrac{E_1}{E_2}$

Hence $\dfrac{2}{7} = \dfrac{240}{E_2}$

Thus the secondary voltage $E_2 = \dfrac{(240)(7)}{(2)} = 840$ V

Further problems on transformers may be found in section C(c), Problems 31 to 34, page 77.

(f) ENERGY STORED IN AN INDUCTOR

Problem 24 An 8 H inductor has a current of 3 A flowing through it. How much energy is stored in the magnetic field of the inductor?

Energy stored, $W = \dfrac{1}{2}LI^2 = \dfrac{1}{2}(8)(3)^2 = 36$ J

Problem 25 A flux of 25 mWb links with a 1500 turn coil when a current of 3 A passes through the coil. Calculate (a) the inductance of the coil, (b) the energy stored in the magnetic field, and (c) the average emf induced if the current falls to zero in 150 ms.

(a) Inductance $L = \dfrac{N\Phi}{I} = \dfrac{(1500)(25 \times 10^{-3})}{3} = 12.5$ H

(b) Energy stored in field $W = \dfrac{1}{2}LI^2 = \dfrac{1}{2}(12.5)(3)^2 = 56.25$ J

(c) Induced emf $E = N\left(\dfrac{\Delta\Phi}{t}\right) = (1500)\left(\dfrac{25 \times 10^{-3}}{150 \times 10^{-3}}\right) = 250$ V

Further problems on energy stored in an inductor may be found in section C(c), Problems 35 to 38, page 77.

C. FURTHER PROBLEMS ON ELECTROMAGNETIC INDUCTION

(a) SHORT ANSWER PROBLEMS

1. What is electromagnetic induction?
2. State Faraday's Laws of electromagnetic induction.
3. State Lenz's Law.
4. Briefly explain the principle of the generator.
5. The direction of an induced emf in a generator may be determined using Fleming's .... rule.
6. To calculate the force on a current-carrying conductor in a magnetic field it is necessary to know the value of three quantities. Name these quantities and their units.
7. The direction of the force on a conductor in a magnetic field may be predetermined using two methods. State each method.
8. Explain briefly the motor principle in terms of the interaction between two magnetic fields.
9. The direction of the magnetic field around a current-carrying conductor is given by the .... rule.
10. The emf E induced in a moving conductor may be calculated using the formula $E = Blv$. Name the quantities represented and their units.
11. What is self inductance? State its symbol.
12. State and define the unit of inductance.
13. When a circuit has an inductance L and the current changes at a rate of $(\Delta I/t)$ then the induced emf E is given by $E = $ .... volts.
14. If a current of I A flowing in a coil of N turns produces a flux of Φ Wb, the coil inductance L is given by $L = $ .... henry's.
15. What is mutual inductance? State its symbol.
16. The mutual inductance between two coils is M. The emf E_2 induced in one coil by a current changing at $(\Delta I_1/t)$ in the other is given by $E_2 = $ .... volts.
17. Briefly explain how a voltage is induced in the secondary winding of a transformer.
18. Draw the circuit diagram symbol for a transformer.
19. State the relationship between turns and voltage ratios for a transformer.
20. The energy W stored by an inductor is given by $W = $ .... joules.

(b) MULTI-CHOICE PROBLEMS (Answers on page 152)

1. A current changing at a rate of 5 A/s in a coil of inductance 5 H induces and emf of
   (a) 25 V in the same direction as the applied voltage;
   (b) 1 V in the same direction as the applied voltage;
   (c) 25 V in the opposite direction to the applied voltage;
   (d) 1 V in the opposite direction to the applied voltage.
2. When a magnetic flux of 10 Wb links with a circuit of 20 turns in 2 s, the induced emf is
   (a) 1 V; (b) 4 V; (c) 100 V; (d) 400 V.
3. A current of 10 A in a coil of 1000 turns produces a flux of 10 mWb linking with the coil. The coil inductance is
   (a) 10^6 H; (b) 1 H; (c) 1 μH; (d) 1 mH.
4. A conductor carries a current of 10 A at right angles to a magnetic field having a flux density of 500 mT. If the length of the conductor in the field is 20 cm the force on the conductor is
   (a) 100 kN; (b) 1 kN; (c) 100 N; (d) 1 N.

5 If a conductor is horizontal, the current flowing from left to right and the direction of the surrounding magnetic field is from above to below, the force exerted on the conductor is
   (a) from left to right;
   (b) from below to above;
   (c) away from the viewer;
   (d) towards the viewer.

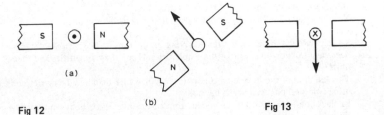

Fig 12

Fig 13

6 For the current-carrying conductor lying in the magnetic field shown in *Fig 12(a)* the direction of the force on the conductor is (a) to the left, (b) upwards, (c) to the right, (d) downwards.

7 For the current-carrying conductor lying in the magnetic field shown in *Fig 12(b)* the direction of the current in the conductor is (a) towards the viewer, (b) away from the viewer.

8 An emf of 1 V is induced in a conductor moving at 10 cm/s in a magnetic field of 0.5 T. The effective length of the conductor in the magnetic field is (a) 20 cm; (b) 5 m; (c) 20 m; (d) 50 m.

9 Which of the following statements is false?
   (a) Fleming's left-hand rule or Lenz's Law may be used to determine the direction of the induced emf.
   (b) An induced emf is set up whenever the magnetic field linking that circuit changes.
   (c) The direction of an induced emf is always such as to oppose the effect producing it.
   (d) The induced emf in any circuit is proportional to the rate of change of the magnetic flux linking the circuit.

10 The mutual inductance between two coils, when a current changing at 20 A/s in one coil induces an emf of 10 mV in the other is
   (a) 0.5 H; (b) 200 mH; (c) 0.5 mH; (d) 2 H.

11 A transformer has 800 primary turns and 100 secondary turns. To obtain 40 V from the secondary winding the voltage applied to the primary winding must be:
   (a) 5 V; (b) 320 V; (c) 2.5 V; (d) 20 V.

12 An emf is induced into a conductor in the direction shown in *Fig 13* when the conductor is moved at a uniform speed in the field between the two magnets. The polarity of the system is
   (a) North pole on right, South pole on left;
   (b) North pole on left, South pole on right.

(c) CONVENTIONAL PROBLEMS

Determination of the force and direction on a current-carrying conductor in a magnetic field

1. A conductor carries a current of 70 A at right angles to a magnetic field having a flux density of 1.5 T. If the length of the conductor in the field is 200 mm calculate the force acting on the conductor. What is the force when the conductor and field are at an angle of 45°?

    [21.0 N; 14.8 N]

2. Calculate the current required in a 240 mm length of conductor of a dc motor when the conductor is situated at right angles to the magnetic field of flux density 1.25 T, if a force of 1.20 N is to be exerted on the conductor.

    [4.0 A]

3. A conductor 30 cm long is situated at right angles to a magnetic field. Calculate the strength of the magnetic field if a current of 15 A in the conductor produces a force on it of 3.6 N.

    [0.80 T]

4. A conductor 300 mm long carries a current of 13 A and is at right angles to a magnetic field between two circular pole faces, each of diameter 80 mm. If the total flux between the pole faces is 0.75 mWb calculate the force exerted on the conductor.

    [0.582 N]

5. (a) A 400 mm length of conductor carrying a current of 25 A is situated at right angles to a magnetic field between two poles of an electric motor. The poles have a circular cross-section. If the force exerted on the conductor is 80 N and the total flux between the pole faces is 1.27 mWb determine the diameter of a pole face.
   (b) If the conductor in part (a) is vertical, the current flowing downwards and the direction of the magnetic field is from left to right, what is the direction of the 80 N force?

    [(a) 14.2 mm; (b) towards the viewer]

Magnitude and directions of induced emf's

6. Find the emf induced in a coil of 200 turns when there is a change of flux of 30 mWb linking with it in 40 ms.

    [150 V]

7. An emf of 25 V is induced in a coil of 300 turns when the flux linking with it changes by 12 mWb. Find the time, in ms, in which the flux makes the change.

    [144 ms]

8. An ignition coil having 10 000 turns has an emf of 8 kV induced in it. What rate of change of flux is required for this to happen?

    [0.8 Wb/s]

9. A flux of 0.35 mWb passing through a 125 turn coil is reversed in 25 ms. Find the average emf induced.

    [3.5 V]

10. Calculate the emf induced in a coil of inductance 6 H by a current changing at a rate of 15 A/s.

    [90 V]

11. An emf of 2 kV is induced in a coil when a current of 5 A collapses uniformly to zero in 10 ms. Determine the inductance of the coil.

    [4 H]

12 An average emf of 50 V is induced in a coil of inductance 160 mH when a current of 7.5 A is reversed. Calculate the time taken for the current to reverse.

[48 ms]

13 A coil of 2500 turns has a flux of 10 mWb linking with it when carrying a current of 2 A. Calculate the coil inductance and the emf induced in the coil when the current collapses to zero in 20 ms.

[12.5 H; 1.25 kV]

14 A conductor of length 15 cm is moved at 750 mm/s at right angles to a uniform flux density of 1.2 T. Determine the emf induced in the conductor.

[0.135 V]

15 Find the speed that a conductor of length 120 mm must be moved at right angles to a magnetic field of flux density 0.6 T to induce in it an emf of 1.8 V.

[25 m/s]

16 A 25 cm long conductor moves at a uniform speed of 8 m/s through a uniform magnetic field of flux density 1.2 T. Determine the current flowing in the conductor when (a) its ends are open-circuited, and (b) its ends are connected to a load of 15 Ω resistance.

[(a) 0; (b) 0.16 A]

17 A straight conductor 500 mm long is moved with constant velocity at right angles both to its length and to a uniform magnetic field. Given that the emf induced in the conductor is 2.5 V and the velocity is 5 m/s, calculate the flux density of the magnetic field. If the conductor forms part of a closed circuit of total resistance 5 ohms, calculate the force on the conductor.

[1 T; 0.25 N]

18 A car is travelling at 80 km/h. Assuming the back axle of the car is 1.76 m in length and the vertical component of the earth's magnetic field is 40 μT, find the emf generated in the axle due to motion.

[1.56 mV]

19 A conductor moves with a velocity of 20 m/s at an angle of (a) 90°; (b) 45°; (c) 30° to a magnetic field produced between two square faced poles of side length 2.5 cm. If the flux on the pole face is 60 mWb find the magnitude of the induced emf in each case.

[(a) 48 V; (b) 33.9 V; (c) 24 V.]

Inductance
20 Calculate the coil inductance when a current of 5 A in a coil of 1000 turns produces a flux of 8 mWb linking with the coil.

[1.6 H]

21 A coil is wound with 600 turns and has a self inductance of 2.5 H. What current must flow to set up a flux of 20 mWb?

[4.8 A]

22 When a current of 2 A flows in a coil, the flux linking with the coil is 80 μWb. If the coil inductance is 0.5 H calculate the number of turns of the coil.

[12 500]

23 A coil of 1200 turns has a flux of 15 mWb linking with it when carrying a current of 4 A. Calculate the coil inductance and the emf induced in the coil when the current collapses to zero in 25 m/s.

[4.5 H; 720 V]

24 A coil has 300 turns and an inductance of 4.5 mH. How many turns would be needed to produce a 0.72 mH coil assuming the same core is used?

[120 turns]

25 A steady current of 5 A when flowing in a coil of 1000 turns produces a magnetic flux of 500 μWb. Calculate the inductance of the coil. The current of 5 A is then reversed in 12.5 ms. Calculate the emf induced in the coil.

[0.1 H; 80 V]

Mutual inductance

26 The mutual inductance between two coils is 150 mH. Find the emf induced in one coil when the current in the other is increasing at the rate of 30 A/s.

[4.5 V]

27 Determine the mutual inductance between two coils when a current changing at 50 A/s in one coil induces an emf of 80 mV in the other.

[1.6 mH]

28 Two coils have a mutual inductance of 0.75 H. Calculate the emf induced in one coil when a current of 2.5 A in the other coil is reversed in 15 ms.

[250 V]

29 The mutual inductance between two coils is 240 mH. If the current in one coil changes from 15 A to 6 A in 12 ms calculate (a) the average emf induced in the other coil and (b) the change of flux linked with the other coil if it is wound with 400 turns.

[(a) 180 V; (b) 5.4 mWb]

30 A mutual inductance of 0.6 H exists between two coils. If a current of 6 A in one coil is reversed in 0.8 s calculate (a) the average emf induced in the other coil and (b) the number of turns on the other coil if the flux change linking with the other coil is 5 mWb.

[(a) 0.9 V; (b) 144]

The transformer

31 A transformer has 800 primary turns and 2000 secondary turns. If the primary voltage is 160 V determine the secondary voltage assuming an ideal transformer.

[400 V]

32 An ideal transformer with a turns ratio of 3:8 is fed from a 240 V supply. Determine its output voltage.

[640 V]

33 An ideal transformer has a turns ratio of 12:1 and is supplied at 192 V. Calculate the secondary voltage.

[16 V]

34 A transformer primary winding connected across a 415 V supply has 750 turns. Determine how many turns must be wound on the secondary side if an output of 1.66 kV is required.

[3000 turns]

Energy stored in an inductor

35 An inductor of 20 H has a current of 2.5 A flowing in it. Find the energy stored in the magnetic field of the inductor.

[62.5 J]

36 Calculate the value of the energy stored when a current of 30 mA is flowing in a coil of inductance 400 mH.

[0.18 mJ]

37 The energy stored in the magnetic field of an inductor is 80 J when the current flowing in the inductor is 2 A. Calculate the inductance of the coil.

[40 H]

38. A flux of 30 mWb links with a 1200 turn coil when a current of 5 A is passing through the coil. Calculate (a) the inductance of the coil, (b) the energy stored in the magnetic field, and (c) the average emf induced if the current is reduced to zero in 0.20 s.

[(a) 7.2 H; (b) 90 J; (c) 180 V]

6 Alternating voltages and currents

A. FORMULAE AND DEFINITIONS ASSOCIATED WITH ALTERNATING VOLTAGES AND CURRENTS

1 Electricity is produced by generators at power stations and then distributed by a vast network of transmission lines (called the National Grid system) to industry and for domestic use. It is easier and cheaper to generate **alternating current** (ac) than direct current (dc) and ac is more conveniently distributed than dc since its voltage can be readily altered using transformers. Whenever dc is needed in preference to ac, devices called rectifiers are used for conversion (see paragraphs 19 to 23 page 83).

2 Let a single turn coil be free to rotate at constant angular velocity ω symmetrically between the poles of a magnet system as shown in *Fig 1*. An emf is generated in coil (from Faraday's Law) which varies in magnitude and reversed its direction at regular intervals. The reason for this is shown in *Fig 2*.

In positions (a), (e) and (i) the conductors of the loop are effectively moving along the magnetic field, no flux is cut and hence no emf is induced. In position (c) maximum flux is cut and hence maximum emf is induced. In position (g), maximum flux is cut and hence maximum emf is again induced.

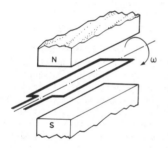

Fig 1 (above)

Fig 2 (right)

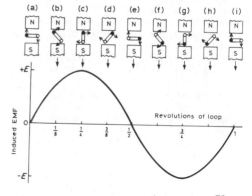

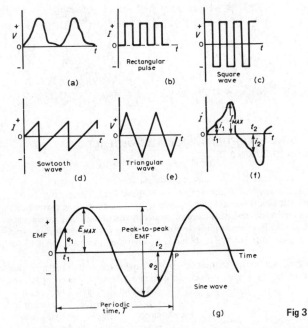

Fig 3

However, using Fleming's right-hand rule, the induced emf is in the opposite direction to that in position (c) and is thus shown as $-E$. In positions (b), (d), (f) and (h) some flux is cut and hence some emf is induced. If all such positions of the coil are considered, in one revolution of the coil, one cycle of alternating emf is produced as shown. This is the principle of operation of the **ac generator** (i.e. the **alternator**).

3 If values of quantities which vary with time t are plotted to a base of time, the resulting graph is called a **waveform**. Some typical waveforms are shown in *Fig 3*. Waveforms (a) and (b) are **unidirectional waveforms**, for, although they vary considerably with time, they flow in one direction only (i.e. they do not cross the time axis and become negative). Waveforms (c) to (g) are called **alternating waveforms** since their quantities are continually changing in direction (i.e. alternately positive and negative).

4 A waveform of the type shown in *Fig 3(g)* is called a **sine wave**. It is the shape of the waveform of emf produced by an alternator and thus the mains electricity supply is of 'sinusoidal' form.

5 One complete series of values is called a **cycle** (i.e. from 0 to P in *Fig 3(g)*).

6 The time taken for an alternating quantity to complete one cycle is called the **period** or the **periodic time**, T, of the waveform.

7 The number of cycles completed in one second is called the **frequency**, f, of the supply and is measured in **hertz**, Hz. The standard frequency of the electricity supply in Great Britain is 50 Hz.

$$T = \frac{1}{f} \quad \text{or} \quad f = \frac{1}{T}.$$

8 **Instantaneous values** are the values of the alternating quantities at any instant of time. They are represented by small letters, i, v, e etc., (see *Figs 3(f)* and *(g)*).

9 The largest value reached in a half cycle is called the **peak value** or the **maximum value** or the **crest value** or the **amplitude** of the waveform. Such values are represented by V_{MAX}, I_{MAX} etc. (see *Figs 3(f)* and *(g)*). A **peak-to-peak** value of emf is shown in *Fig 3(g)* and is the difference between the maximum and minimum values in a cycle.

10 The **average or mean value** of a symmetrical alternating quantity, (such as a sine wave), is the average value measured over a half cycle, (since over a complete cycle the average value is zero).

$$\text{Average or mean value} = \frac{\text{area under the curve}}{\text{length of base}}$$

The area under the curve is found by approximate methods such as the trapezoidal rule, the mid-ordinate rule or Simpson's rule. Average values are represented by V_{AV}, I_{AV}, etc.

For a sine wave, average value = $0.637 \times$ maximum value (i.e. $2/\pi \times$ maximum value).

11 The **effective value** of an alternating current is that current which will produce the same heating effect as an equivalent direct current. The effective value is called the **root mean square (rms) value** and whenever an alternating quantity is given, it is assumed to be the rms value. For example, the domestic mains supply in Great Britain is 240 V and is assumed to mean '240 V rms'. The symbols used for rms values are I, V, E, etc. For a non-sinusoidal waveform as shown in *Fig 4* the rms value is given by:

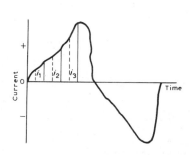

$$I = \sqrt{\left[\frac{i_1^2 + i_2^2 + \ldots + i_n^2}{n}\right]}$$

where n is the number of intervals used.

For a sine wave, rms value = $0.707 \times$ maximum value (i.e. $1/\sqrt{2} \times$ maximum value).

Fig 4

12 (a) Form factor = $\dfrac{\text{rms value}}{\text{average value}}$. For a sine wave, form factor = 1.11.

(b) Peak factor = $\dfrac{\text{maximum value}}{\text{rms value}}$. For a sine wave, peak factor = 1.41.

The values of form and peak factors give an indication of the shape of waveforms.

13 In *Fig 5*, OA represents a vector that is free to rotate anticlockwise about 0 at an angular velocity of ω rad/s. A rotating vector is known as a **phasor**. After time t seconds the vector OA has turned through an angle ωt. If the line BC is constructed perpendicular to OA as shown, then

$$\sin \omega t = \frac{BC}{OB} \quad \text{i.e. } BC = OB \sin \omega t.$$

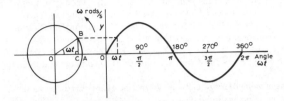

Fig 5

If all such vertical components are projected on to a graph of y against angle ωt (in radians), a sine curve results of maximum value OA. Any quantity which varies sinusoidally can thus be represented as a phasor.

14 A sine curve may not always start at 0°. To show this a periodic function is represented by $y = \sin(\omega t \pm \phi)$, where ϕ is a phase (or angle) difference compared

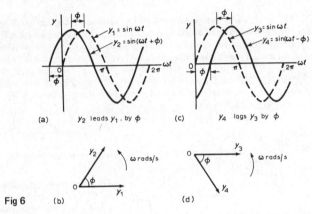

Fig 6

with $y = \sin \omega t$. In *Fig 6(a)*, $y_2 = \sin(\omega t + \phi)$ starts ϕ radians earlier than $y_1 = \sin \omega t$ and is thus said to **lead** y_1 by ϕ radians. Phasors y_1 and y_2 are shown in *Fig 6(b)* at the time when $t = 0$. In *Fig 6(c)*, $y_4 = \sin(\omega t - \phi)$ starts ϕ radians later than $y_3 = \sin \omega t$ and is thus said to **lag** y_3 by ϕ radians. Phasors y_3 and y_4 are shown in *Fig 6(d)* at the time when $t = 0$.

15 Given the general sinusoidal voltage, $v = V_{MAX} \sin(\omega t \pm \phi)$, then
  (i) Amplitude of maximum value $= V_{MAX}$
  (ii) Peak to peak value $= 2V_{MAX}$
  (iii) Angular velocity $= \omega$ rads/s
  (iv) Periodic time, $T = 2\pi/\omega$ seconds.
  (v) Frequency, $f = \omega/2\pi$ Hz (hence $\omega = 2\pi f$).
  (vi) ϕ = angle of lag or lead (compared with $v = V_{MAX} \sin \omega t$).

16 The resultant of the addition (or subtraction) of two sinusoidal quantities may be determined either:
  (a) by plotting the periodic functions graphically (see worked *Problems 13 and 16*), or
  (b) by resolution of phasors by drawing or calculation (see worked *Problems 14 and 15*).

17 When a sinusoidal voltage is applied to a purely resistive circuit of resistance R, the voltage and current waveforms are in phase and $I = V/R$ (exactly as in a dc circuit). V and I are rms values.
18 For an ac resistive circuit, power $P = VI = I^2R = V^2/R$ watts (exactly as in a dc circuit). V and I are rms values.
19 The process of obtaining unidirectional currents and voltages from alternating currents and voltages is called **rectification**. Automatic switching in circuits is carried out by devices called diodes (see Chapter 8).
20 Using a single diode, as shown in *Fig 7*, **half-wave rectification** is obtained. When P is sufficiently positive with respect to Q, diode D is switched on and current i flows. When P is negative with respect to Q, diode D is switched off. Transformer T isolates the equipment from direct connection with the mains supply and enables the mains voltage to be changed.

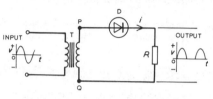

Fig 7

21 Two diodes may be used as shown in *Fig 8* to obtain **full wave rectification**. A centre-tapped transformer T is used. When P is sufficiently positive with respect to Q, diode D_1 conducts and current flows (shown by the broken line in *Fig 8*). When S is positive with respect to Q, diode D_2 conducts and current flows (shown by continuous line in *Fig 8*). The current flowing in R is in the same direction for both half cycles of the input. The output waveform is thus as shown in *Fig 8*.

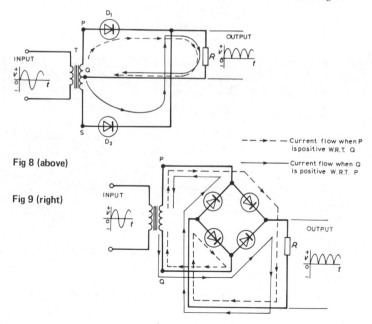

Fig 8 (above)

Fig 9 (right)

– – –▶– – – Current flow when P is positive W.R.T. Q

———▶——— Current flow when Q is positive W.R.T. P

83

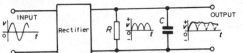

Fig 10

22. Four diodes may be used in a **bridge rectifier** circuit, as shown in *Fig 9* to obtain **full wave rectification**. As for the rectifier shown in *Fig 8*, the current flowing in R is in the same direction for both half cycles of the input giving the output waveform shown.
23. To smooth the output of the rectifiers described above, capacitors having a large capacitance may be connected across the load resistors R. The effect of this is shown on the output in *Fig 10*.

B. WORKED PROBLEMS ON ALTERNATING VOLTAGES AND CURRENTS

(a) FREQUENCY AND PERIODIC TIME

Problem 1 Determine the periodic time for frequencies of (a) 50 Hz and (b) 20 kHz.

(a) Periodic time $T = \dfrac{1}{f} = \dfrac{1}{50} = 0.02$ s or 20 ms

(b) Periodic time $T = \dfrac{1}{f} = \dfrac{1}{20\,000} = 0.000\,05$ s or 50 μs

Problem 2 Determine the frequencies for periodic times of (a) 4 ms, (b) 4 μs.

(a) Frequency $f = \dfrac{1}{T} = \dfrac{1}{4 \times 10^{-3}} = \dfrac{1000}{4} = 250$ Hz

(b) Frequency $f = \dfrac{1}{T} = \dfrac{1}{4 \times 10^{-6}} = \dfrac{1\,000\,000}{4} = 250\,000$ Hz or 250 kHz or 0.25 MHz

Problem 3 An alternating current completes 5 cycles in 8 ms. What is its frequency?

Time for 1 cycle $= \dfrac{8}{5}$ ms $= 1.6$ ms $=$ periodic time T.

Frequency $f = \dfrac{1}{T} = \dfrac{1}{1.6 \times 10^{-3}} = \dfrac{1000}{1.6} = \dfrac{10\,000}{16} = 625$ Hz

Further problems on frequency and periodic time may be found in section C(c), Problems 1 to 3, page 97.

(b) AC VALUES OF NON-SINUSOIDAL WAVEFORMS

> *Problem 4* For the periodic waveforms shown in *Fig 11* determine for each: (i) frequency; (ii) average value over half a cycle; (iii) rms value; (iv) form factor; and (v) peak factor.

(a) *Triangular waveform (Fig 11(a))*

(i) Time for 1 complete cycle = 20 ms = periodic time, T.

Hence frequency $f = \dfrac{1}{T} = \dfrac{1}{20 \times 10^{-3}} = \dfrac{1000}{20} = 50$ Hz

(ii) Area under the triangular waveform for a half cycle

$= \dfrac{1}{2} \times$ base $\times$ height

$= \dfrac{1}{2} \times (10 \times 10^{-3}) \times 200$

$= 1$ volt second.

Average value of waveform

$= \dfrac{\text{area under curve}}{\text{length of base}}$

$= \dfrac{1 \text{ volt second}}{10 \times 10^{-3} \text{ second}}$

$= \dfrac{1000}{10}$

$= \mathbf{100}$ **V**

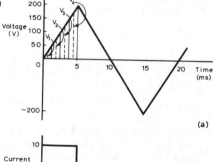

(iii) In *Fig 11(a)*, the first 1/4 cycle is divided into 4 intervals.

Thus rms value

$= \sqrt{\left[\dfrac{i_1{}^2 + i_2{}^2 + i_3{}^2 + i_4{}^2}{4}\right]}$

$= \sqrt{\left[\dfrac{25^2 + 75^2 + 125^2 + 175^2}{4}\right]}$

$= 114.6$ **V**

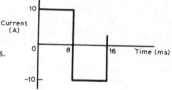

Fig 11

(Note that the greater the number of intervals chosen, the greater the accuracy of the result. For example, if twice the number of ordinates as that chosen above are used, the rms value is found to be 115.6 V)

(iv) Form factor $= \dfrac{\text{rms value}}{\text{average value}} = \dfrac{114.6}{100} = \mathbf{1.15}$

(v) Peak factor $= \dfrac{\text{maximum value}}{\text{rms value}} = \dfrac{200}{114.6} = \mathbf{1.75}$

(b) *Rectangular waveform (Fig 11(b))*

(i) Time for 1 complete cycle = 16 ms = periodic time, T.

Hence frequency, $f = \dfrac{1}{T} = \dfrac{1}{16 \times 10^{-3}} = \dfrac{1000}{16} = 62.5$ Hz

(ii) Average value over half a cycle = $\dfrac{\text{area under curve}}{\text{length of base}}$

$$= \dfrac{10 \times (8 \times 10^{-3})}{8 \times 10^{-3}} = 10 \text{ A}$$

(iii) The rms value = $\sqrt{\left[\dfrac{i_1^2 + i_2^2 + \ldots + i_n^2}{n}\right]} = 10$ A,

however many intervals are chosen, since the waveform is rectangular.

(iv) Form factor = $\dfrac{\text{rms value}}{\text{average value}} = \dfrac{10}{10} = 1$

(v) Peak factor = $\dfrac{\text{maximum value}}{\text{rms value}} = \dfrac{10}{10} = 1$

Problem 5 The following table gives the corresponding values of current and time for a half cycle of alternating current.

time t (ms)	0	0.5	1.0	1.5	2.0	2.5	3.0	3.5	4.0	4.5	5.0
current i (A)	0	7	14	23	40	56	68	76	60	5	0

Assuming the negative half cycle is identical in shape to the positive half cycle, plot the waveform and find (a) the frequency of the supply, (b) the instantaneous values of current after 1.25 ms and 3.8 ms, (c) the peak or maximum value, (d) the mean or average value, and (e) the rms value of the waveform.

The half cycle of alternating current is shown plotted in *Fig 12*.

(a) Time for a half cycle = 5 ms. Hence the time for 1 cycle, i.e. the periodic time, $T = 10$ ms or 0.01 s.

Frequency, $f = \dfrac{1}{T} = \dfrac{1}{0.01} = 100$ Hz.

(b) Instantaneous value of current after 1.25 ms is **19 A**, from *Fig 12*.
Instantaneous value of current after 3.8 ms is **70 A**, from *Fig 12*.

(c) Peak or maximum value = **76A**

(d) Mean or average value = $\dfrac{\text{area under curve}}{\text{length of base}}$

Using the mid-ordinate rule with 10 intervals, each of width 0.5 ms gives:

Area under curve = $(0.5 \times 10^{-3})[3+10+19+30+49+63+73+72+30+2]$
(see *Fig 12*)

$= (0.5 \times 10^{-3})(351)$

Hence mean or average value = $\dfrac{(0.5 \times 10^{-3})(351)}{5 \times 10^{-3}} = $ **35.1 A**

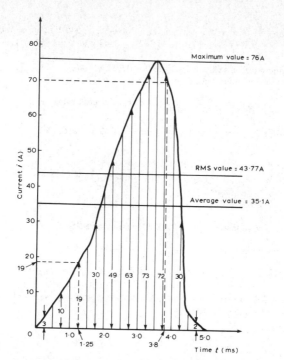

Fig 12

(e) rms value = $\sqrt{\left[\dfrac{3^2+10^2+19^2+30^2+49^2+63^2+73^2+72^2+30^2+2^2}{10}\right]}$

$= \sqrt{\left[\dfrac{19\,157}{10}\right]} = $ **43.8 A**

Further problems on ac values of non-sinusoidal waveforms may be found in section C(c), Problems 4 to 7, page 97.

(c) AC VALUES OF SINUSOIDAL WAVEFORMS

Problem 6 Calculate the rms value of a sinusoidal current of maximum value 20 A.

For a sine wave,
rms value = 0.707 × maximum value
= 0.707 × 20 = **14.14 A**

Problem 7 Determine the peak and mean values for a 240 V mains supply.

For a sine wave, rms value of voltage $V = 0.707 \times V_{MAX}$
A 240 V mains supply means that 240 V is the rms value.

Hence $V_{MAX} = \dfrac{V}{0.707} = \dfrac{240}{0.707} = 339.5$ V = **peak value**

Mean value $V_{AV} = 0.637\, V_{MAX} = 0.637 \times 339.5 = \mathbf{216.3\ V}$

Problem 8 A supply voltage has a mean value of 150 V. Determine its maximum value and its rms value.

For a sine wave, mean value = 0.637 × maximum value.

Hence maximum value = $\dfrac{\text{mean value}}{0.637} = \dfrac{150}{0.637} = \mathbf{235.5\ V}$

rms value = 0.707 × maximum value = 0.707 × 235.5 = **166.5 V**

Further problems on ac values of sinusoidal waveforms may be found in section C(c), Problems 8 to 12, page 98.

(d) $v = V_{MAX} \sin(\omega t \pm \phi)$

Problem 9 An alternating voltage is given by $v = 282.8 \sin 314\,t$ volts. Find (a) the rms voltage, (b) the frequency and (c) the instantaneous value of voltage when $t = 4$ ms.

(a) The general expression for an alternating voltage is $v = V_{MAX} \sin(\omega t \pm \phi)$.
   Comparing $v = 282.8 \sin 314\,t$ with this general expression gives the peak voltage as 282.8 V
   Hence the rms voltage = 0.707 × maximum value = 0.707 × 282.8 = **200 V**

(b) Angular velocity, $\omega = 314$ rads/s i.e. $2\pi f = 314$

   Hence frequency, $f = \dfrac{314}{2\pi} = \mathbf{50\ Hz}$

(c) When $t = 4$ ms, $v = 282.8 \sin(314 \times 4 \times 10^{-3})$
   $= 282.8 \sin(1.256)$

   $1.256\ \text{radians} = \left(1.256 \times \dfrac{180}{\pi}\right)^\circ = 71.96^\circ = 71^\circ\,58'$

   Hence $v = 282.8 \sin 71^\circ\,58'$
   $= \mathbf{268.9\ V}$

Problem 10 An alternating voltage is given by $v = 75 \sin(200\pi t - 0.25)$ volts. Find (a) the amplitude; (b) the peak-to-peak value; (c) the rms value, (d) the periodic time; (e) the frequency; and (f) the phase angle (in degrees and minutes) relative to $75 \sin 200\pi t$.

Comparing $v = 75 \sin(200\pi t - 0.25)$ with the general expression
$v = V_{MAX} \sin(\omega t \pm \phi)$ gives:
(a) Amplitude, or peak value = **75 V**
(b) Peak-to-peak value = $2 \times 75 = $ **150 V**
(c) The rms value = $0.707 \times$ maximum value = $0.707 \times 75 = $ **53 V**
(d) Angular velocity, $\omega = 200\pi$ rads/s

Hence periodic time, $T = \dfrac{2\pi}{\omega} = \dfrac{2\pi}{200\pi} = \dfrac{1}{100} = $ **0.01 s or 10 ms**

(e) Frequency, $f = \dfrac{1}{T} = \dfrac{1}{0.01} = $ **100 Hz**

(f) Phase angle, $\phi = 0.25$ radians lagging $75 \sin 200\pi t$

$$0.25 \text{ rads} = \left(0.25 \times \dfrac{180}{\pi}\right)^\circ = 14.32° = 14° \; 19'$$

Hence phase angle = **14° 19' lagging**

Problem 11 An alternating voltage, v, has a periodic time of 0.01 s and a peak value of 40 V. When time t is zero, $v = -20$ V. Express the instantaneous voltage in the form $v = V_{MAX} \sin(\omega t + \phi)$ V.

Amplitude, $V_{MAX} = 40$ V.
Periodic time $T = \dfrac{2\pi}{\omega}$. Hence angular velocity, $\omega = \dfrac{2\pi}{T} = \dfrac{2\pi}{0.01} = 200\pi$ rads/s
$v = V_{MAX} \sin(\omega t + \phi)$ thus becomes $v = 40 \sin(200\pi t + \phi)$ V.
When time $t = 0, v = -20$ V
i.e. $-20 = 40 \sin \phi$

$\sin \phi = \dfrac{-20}{40} = -0.5$

Hence $\phi = \arcsin -0.5 = -30° = \left(-30 \times \dfrac{\pi}{180}\right)$ rads $= -\dfrac{\pi}{6}$ rads

Thus $v = 40 \sin\left(200\pi t - \dfrac{\pi}{6}\right)$ V

Problem 12 The current in an ac circuit at any time t seconds is given by:
$i = 120 \sin(100\pi t + 0.36)$ amperes. Find:
(a) the peak value, the periodic time, the frequency and phase angle relative to $120 \sin 100\pi t$;
(b) the value of the current when $t = 0$;
(c) the value of the current when $t = 8$ ms;
(d) the time when the current first reaches 60 A, and
(e) the time when the current is first a maximum.

(a) Peak value = **120 A**.

Periodic time $T = \dfrac{2\pi}{\omega} = \dfrac{2\pi}{100\pi}$ (since $\omega = 100\pi$)

$= \dfrac{1}{50} = $ **0.02 s or 20 ms**

Frequency $f = \dfrac{1}{T} = \dfrac{1}{0.02} = $ **50 Hz**

Phase angle = 0.36 rads = $\left(0.36 \times \dfrac{180}{\pi}\right)^{\circ} = $ **20° 38′ leading**

(b) When $t = 0$, $i = 120 \sin(0 + 0.36) = 120 \sin 20° 38' = $ **42.29 A**

(c) When $t = 8$ ms, $i = 120 \sin\left[100\pi \left(\dfrac{8}{10^3}\right) + 0.36\right]$

$= 120 \sin 2.8733$
$= 120 \sin 164° 38'$
$= $ **31.80 A**

(d) When $i = 60$ A, $60 = 120 \sin(100\pi t + 0.36)$

$\dfrac{60}{120} = \sin(100\pi t + 0.36)$

$(100\pi t + 0.36) = \arcsin 0.5 = 30° = \dfrac{\pi}{6}$ rads = 0.5236 rads

Hence time, $t = \dfrac{0.5236 - 0.36}{100\pi} = $ **0.5208 ms**

(e) When the current is a maximum, $i = 120$ A

Thus $120 = 120 \sin(100\pi t + 0.36)$

$1 = \sin(100\pi t + 0.36)$

$(100\pi t + 0.36) = \arcsin 1 = 90° = \dfrac{\pi}{2}$ rads = 1.5708 rads

Hence time, $t = \dfrac{1.5708 - 0.36}{100\pi} = $ **3.854 ms**

Further problems on the general expression $v = V_{MAX} \sin(\omega t \pm \phi)$ may be found in section C(c), Problems 13 to 18, page 98.

(e) COMBINATION OF PERIODIC FUNCTIONS

Problem 13 The instantaneous values of two alternating currents are given by $i_1 = 20 \sin \omega t$ amperes and $i_2 = 10 \sin(\omega t + \pi/3)$ amperes. By plotting i_1 and i_2 on the same axes, using the same scale, over one cycle, and adding ordinates at intervals, obtain a sinusoidal expression for $i_1 + i_2$.

$i_1 = 20 \sin \omega t$ and $i_2 = 10 \sin\left(\omega t + \dfrac{\pi}{3}\right)$ are shown plotted in *Fig 13*.

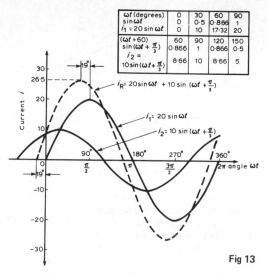

Fig 13

Ordinates of i_1 and i_2 are added at, say, 15° intervals (a pair of dividers are useful for this).

For example

at 30°, $i_1+i_2 = 10+10 = 20$ A,
at 60°, $i_1+i_2 = 8.7+17.3 = 26$ A,
at 150°, $i_1+i_2 = 10+(-5) = 5$ A, and so on.

The resultant waveform for i_1+i_2 is shown by the broken line in *Fig 13*. It has the same period, and hence frequency, as i_1 and i_2. The amplitude or peak value is 26.5 A. The resultant waveform leads the curve $i_1 = 20 \sin \omega t$ by 19°

i.e. $\left(19 \times \dfrac{\pi}{180}\right)$ rads = 0.332 rads

Hence the sinusoidal expression for the resultant i_1+i_2 is given by:

$i_R = i_1+i_2 = 26.5 \sin(\omega t+0.332)$ A

Problem 14 Two alternating voltages are represented by $v_1 = 50 \sin \omega t$ volts and $v_2 = 100 \sin(\omega t - \pi/6)$ V. Draw the phasor diagram and find, by calculation, a sinusoidal expression to represent v_1+v_2.

Phasors are usually drawn at the instant when time $t = 0$. Thus v_1 is drawn horizontally 50 units long and v_2 is drawn 100 units long lagging v_1 by $\pi/6$ rads, i.e. 30°. This is shown in *Fig 14(a)* where 0 is the point of rotation of the phasors. Procedure to draw phasor diagram to represent v_1+v_2:

(i) Draw v_1 horizontal 50 units long, i.e. oa of *Fig 14(b)*.
(ii) Join v_2 to the end of v_1 at the appropriate angle, i.e. ab of *Fig 14(b)*.
(iii) The resultant $v_R = v_1+v_2$ is given by the length ob and its phase angle may be measured with respect to v_1

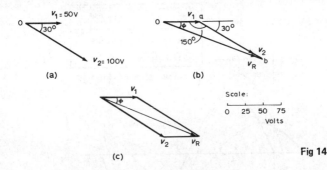

Fig 14

Alternatively, when two phasors are being added the resultant is always the diagonal of the parallelogram, as shown in *Fig 14(c)*.

From the drawing, by measurement, $v_R = 145$ V and angle $\phi = 20°$ lagging v_1. A more accurate solution is obtained by calculation, using the cosine and sine rules. Using the cosine rule on triangle oab of *Fig 14(b)* gives:

$$\begin{aligned} v_R^2 &= v_1^2 + v_2^2 - 2v_1 v_2 \cos 150° \\ &= 50^2 + 100^2 - 2(50)(100) \cos 150° \\ &= 2500 + 10\,000 - (-8660) \\ v_R &= \sqrt{(21\,160)} = 145.5 \text{ V}. \end{aligned}$$

Using the sine rule, $\dfrac{100}{\sin \phi} = \dfrac{145.5}{\sin 150°}$

$\sin \phi = \dfrac{100 \sin 150°}{145.5} = 0.3436$

$\phi = \arcsin 0.3436 = 20° \, 6' = 0.35$ radians, and lags v_1

Hence $v_R = v_1 + v_2 = 145.5 \sin (\omega t - 0.35)$ V

Problem 15 Find a sinusoidal expression for $(i_1 + i_2)$ of *Problem 13*, (a) by drawing phasors, (b) by calculation.

(a) The relative positions of i_1 and i_2 at time $t = 0$ are shown as phasors in *Fig 15(a)*. The phasor diagram in *Fig 15(b)* shows the resultant i_R, and i_R is measured as 26 A and angle ϕ as 19° (i.e. 0.33 rads) leading i_1.

Hence, by drawing,
$i_R = 26 \sin (\omega t + 0.33)$ A

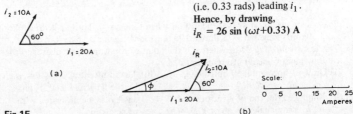

Fig 15

(b) From *Fig 15(b)*, by the cosine rule:

$i_R^2 = 20^2 + 10^2 - 2(20)(10)(\cos 120°)$

from which $i_R = 26.46$ A

By the sine rule: $\dfrac{10}{\sin \phi} = \dfrac{26.46}{\sin 120°}$

from which $\phi = 19° \, 10'$ (i.e. 0.333 rads)

Hence, by calculation $i_R = 26.46 \sin(\omega t + 0.333)$ A

Problem 16 Two alternating voltages are given by $v_1 = 120 \sin \omega t$ volts and $v_2 = 200 \sin(\omega t - \pi/4)$ volts. Obtain sinusoidal expressions for $v_1 - v_2$ (a) by plotting waveforms, and (b) by resolution of phasors.

(a) $v_1 = 120 \sin \omega t$ and $v_2 = 200 \sin(\omega t - \pi/4)$ are shown plotted in *Fig 16*. Care must be taken when subtracting values of ordinates especially when at least one of the ordinates is negative. For example

at 30°, $v_1 - v_2 = 60 - (-52) = 112$ V
at 60°, $v_1 - v_2 = 104 - 52 = 52$ V
at 150°, $v_1 - v_2 = 60 - 193 = -133$ V and so on

The resultant waveform, $v_R = v_1 - v_2$, is shown by the broken line in *Fig 16*. The maximum value of v_R is 143 V and the waveform is seen to lead v_1 by 99° (i.e. 1.73 radians).

Hence, by drawing, $v_R = v_1 - v_2 = 143 \sin(\omega t + 1.73)$ volts

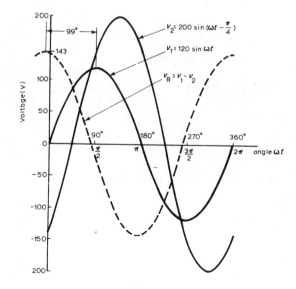

Fig 16

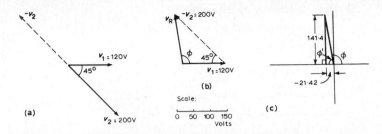

Fig 17

(b) The relative positions of v_1 and v_2 are shown at time $t = 0$ as phasors in *Fig 17(a)*. Since the resultant of $v_1 - v_2$ is required, $-v_2$ is drawn in the opposite direction to $+v_2$ (shown by the broken line in *Fig 17(a)*). The phasor diagram with the resultant is shown in *Fig 17(b)* where $-v_2$ is added phasorially to v_1.

By resolution:
Sum of horizontal components of v_1 and v_2 = $120 \cos 0° - 200 \cos 45° = -21.42$
Sum of vertical components of v_1 and v_2 = $120 \sin 0° + 200 \sin 45° = 141.4$
From *Fig 17(c)*, resultant $v_R = \sqrt{[(-21.42)^2 + (141.4)^2]} = 143.0$,

and $\tan \phi' = \dfrac{141.4}{21.42} = \tan 6.6013$, from which

ϕ' = arctan 6.6013 = 81°23' and
ϕ = 98° 37' or 1.721 radians.

Hence, by resolution of phasors, $v_R = v_1 - v_2 = 143.0 \sin (\omega t + 1.721)$ volts.

Further problems on combinations of periodic functions may be found in section C(c), Problems 19 to 22, page 99.

(f) PURELY RESISTIVE AC CIRCUITS

Problem 17 Determine the current flowing in a $20\,\Omega$ resistor when a 240V, 50Hz supply voltage is applied across the resistor. Find also the power dissipated by the resistor.

For a purely resistive ac circuit

current $I = \dfrac{V}{R} = \dfrac{240}{20} = 12$ A

power $P = VI = 240 \times 12 = 2880$ W
$ = 2.88$ kW

Problem 18 A sinusoidal voltage of maximum value 50 V causes a current of maximum value 4 A to flow through a resistance. Find the value of the resistance and the power developed. What is the energy dissipated in 2 minutes?

rms value of voltage, $V = 0.707 \times$ maximum value $= 0.707 \times 50 = 35.35$ V
rms value of current, $I = 0.707 \times$ maximum value $= 0.707 \times 4 = 2.828$ A

For a purely resistive ac circuit, resistance $R = \dfrac{V}{I} = \dfrac{35.35}{2.828} = 12.5\ \Omega$

Power developed, $P = VI = 35.35 \times 2.828 = $ **100 W**

Energy dissipated $=$ Power $\times$ time $= 100 \times (2 \times 60)$ W s

$= 12\,000$ J $=$ **12 kJ**

Further problems on purely resistive ac circuits may be found in section C(c), Problems 23 to 26, page 99.

C. FURTHER PROBLEMS ON ALTERNATING VOLTAGES AND CURRENTS

(a) SHORT ANSWER PROBLEMS

1 Briefly explain the principle of the simple alternator.
2 What is the difference between an alternating and a unidirectional waveform.
3 What is meant by (a) waveform; (b) cycle.
4 The time to complete one cycle of a waveform is called the ....
5 What is frequency? Name its unit.
6 The mains supply voltage has a special shape of waveform called a ....
7 Define peak value.
8 What is meant by the rms value?
9 The domestic mains electricity supply voltage in Great Britain is ....
10 What is the mean value of a sinusoidal alternating emf which has a maximum value of 100 V?
11 The effective value of a sinusoidal waveform is .... $\times$ maximum value.
12 What is a phasor quantity?
13 Complete the statement: Form factor $= \dfrac{\ldots\ldots\ldots\ldots}{\ldots\ldots\ldots\ldots}$, and for a sine wave, form factor $= \dfrac{\ldots\ldots\ldots\ldots}{\ldots\ldots\ldots\ldots}$.
14 Complete the statement: Peak factor $= \dfrac{\ldots\ldots\ldots\ldots}{\ldots\ldots\ldots\ldots}$, and for a sine wave, peak factor $= \dfrac{\ldots\ldots\ldots\ldots}{\ldots\ldots\ldots\ldots}$.
15 A sinusoidal current is given by $i = I_{MAX} \sin(\omega t \pm \alpha)$. What do the symbols I_{MAX}, ω and α represent?
16 A sinusoidal voltage of 250 V is applied across a pure resistance of 5 Ω. What is (a) the current flowing, and (b) the power developed across the resistance?
17 How is switching obtained when converting ac to dc?
18 Draw an appropriate circuit diagram suitable for half-wave rectification.

19 How may full-wave rectification be achieved?
20 What is a simple method of smoothing the output of a rectifier?

(b) MULTI-CHOICE PROBLEMS (Answers on page 152)

1 The value of an alternating current at any given instant is (a) a maximum value; (b) a peak value; (c) an instantaneous value; (d) an rms value.
2 An alternating current completes 100 cycles in 0.2 s. Its frequency is (a) 20 Hz; (b) 100 Hz; (c) 0.002 Hz; (d) 1 kHz.
3 In *Fig 18*, at the instant shown the generated emf will be (a) zero; (b) an rms value; (c) an average value; (d) a maximum value.
4 The supply of electrical energy for a consumer is usually by ac because
   (a) transmission and distribution are more easily effected;
   (b) it is most suitable for variable speed motors;
   (c) the volt drop in cables is minimal;
   (d) cable power losses are negligible.
5 Which of the following statements is false?
   (a) It is cheaper to use ac than dc.
   (b) Distribution of ac is more convenient than with dc since voltages may be readily altered using transformers.
   (c) An alternator is an ac generator.
   (d) A rectifier changes dc into ac.

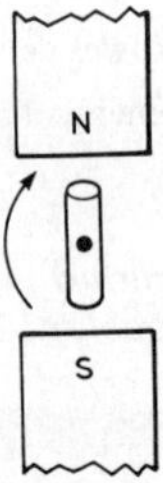

Fig 18

6 An alternating voltage of maximum value 100 V is applied to a lamp. Which of the following direct voltages, if applied to the lamp, would cause the lamp to light with the same brilliance?
   (a) 100 V; (b) 63.7 V; (c) 70.7 V; (d) 141.4 V.
7 The value normally stated when referring to alternating currents and voltages is the (a) instantaneous value; (b) rms value; (c) average value; (d) peak value.
8 State which of the following is false. For a sine wave:
   (a) the peak factor is 1.414.
   (b) the rms value is 0.707 × peak value.
   (c) the average value is 0.637 × rms value.
   (d) the form factor is 1.11.
9 An ac supply is 70.7 V, 50 Hz. Which of the following statements is false?
   (a) The periodic time is 20 ms.
   (b) The peak value of the voltage is 70.7 V.
   (c) The rms value of the voltage is 70.7 V.
   (d) The peak value of the voltage is 100 V.
10 An alternating voltage is given by $v = 100 \sin(50\pi t - 0.30)$ V. Which of the following statements is true?
   (a) The rms voltage is 100 V;
   (b) The periodic time is 20 ms;
   (c) The frequency is 25 Hz;
   (d) The voltage is leading $v = 100 \sin 50\pi t$ by 0.30 radians.

(c) CONVENTIONAL PROBLEMS

Frequency and periodic time

1. Determine the periodic time for the following frequencies:
   (a) 2.5 Hz; (b) 100 Hz; (c) 40 kHz.

   [(a) 0.4 s; (b) 10 ms; (c) 25 μs]

2. Calculate the frequency for the following periodic times:
   (a) 5 ms; (b) 50 μs; (c) 0.2 s.

   [(a) 0.2 kHz; (b) 20 kHz; (c) 5 Hz]

3. An alternating current completes 4 cycles in 5 ms. What is its frequency?

   [800 Hz]

ac values of non-sinusoidal waveforms

4. An alternating current varies with time over half a cycle as follows:

Current (A)	0	0.7	2.0	4.2	8.4	8.2	2.5	1.0	0.4	0.2	0
time (ms)	0	1	2	3	4	5	6	7	8	9	10

   The negative half cycle is similar. Plot the curve and determine:
   (a) the frequency; (b) the instantaneous values at 3.4 ms and 5.8 ms; (c) its mean value and (d) its rms value.

   [(a) 50 Hz. (b) 5.5 A, 3.4 A; (c) 2.8 A; (d) 4.0 A]

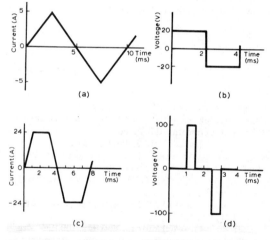

Fig 19

5. For the waveforms shown in *Fig 19* determine for each (i) the frequency; (ii) the average value over half a cycle; (iii) the rms value; (iv) the form factor; (v) the peak factor.

   ⎡(a) (i) 100 Hz; (ii) 2.50A; (iii) 2.88A; (iv) 1.15; (v) 1.74⎤
   ⎢(b) (i) 250 Hz; (ii) 20 V; (iii) 20 V; (iv) 1.0; (v) 1.0⎥
   ⎢(c) (i) 125 Hz; (ii) 18 A; (iii) 19.56 A;(iv) 1.09; (v) 1.23⎥
   ⎣(d) (i) 250 Hz; (ii) 25 V; (iii) 50 V; (iv) 2.0; (v) 2.0⎦

6. An alternating voltage is triangular in shape, rising at a constant rate to a maximum of 300 V in 8 ms and then falling to zero at a constant rate in 4 ms. The negative half cycle is identical in shape to the positive half cycle. Calculate (a) the mean voltage over half a cycle, and (b) the rms voltage.

[(a) 150 V; (b) 106 V]

7. An alternating emf varies with time over half a cycle as follows:

Emf (V)	0	45	80	155	215	320	210	95	0
time (ms)	0	1.5	3.0	4.5	6.0	7.5	9.0	10.5	12.0

The negative half cycle is identical in shape to the positive half cycle. Plot the waveform and determine (a) the periodic time and frequency; (b) the instantaneous value of voltage at 3.75 ms, (c) the times when the voltage is 125 V; (d) the mean value, and (e) the rms value.

[(a) 24 ms, 41.67 Hz; (b) 115 V; (c) 4 ms and 10.1 ms; (d) 142 V; (e) 171 V]

ac values of sinusoidal waveforms

8. Calculate the rms value of a sinusoidal curve of maximum value 300 V.

[212.1 V]

9. Find the peak and mean values for a 200 V mains supply.

[282.9 V; 180.2 V]

10. Plot a sine wave of peak value 10.0 A. Show that the average value of the waveform is 6.37 A over half a cycle, and that the rms value is 7.07 A.

11. A sinusoidal voltage has a maximum value of 120 V. Calculate its rms and average values.

[84.8 V; 76.4 V]

12. A sinusoidal current has a mean value of 15.0 A. Determine its maximum and rms values.

[23.55 A; 16.65 A]

$v = V_{MAX} \sin(\omega t \pm \phi)$

13. An alternating voltage is represented by $v = 20 \sin 157.1 t$ volts. Find (a) the maximum value; (b) the frequency; and (c) the periodic time. (d) What is the angular velocity of the phasor representing this waveform?

[(a) 20 V; (b) 25 Hz; (c) 0.04 s; (d) 157.1 rads/s]

14. Find the peak value, the rms value, the periodic time, the frequency and the phase angle (in degrees and minutes) of the following alternating quantities:
(a) $v = 90 \sin 400\pi t$ volts.   [(a) 90 V; 63.63 V; 5 ms; 200 Hz; 0°]
(b) $i = 50 \sin(100\pi t + 0.30)$ amperes.

[(b) 50 A; 35.35 A; 0.02 s; 50 Hz; 17°11′ lead]

(c) $e = 200 \sin(628.4t - 0.41)$ volts.

[(c) 200 V; 141.4 V; 0.01 s; 100 Hz; 23°29′ lag]

15. A sinusoidal current has a peak value of 30 A and a frequency of 60 Hz. At time $t = 0$, the current is zero. Express the instantaneous current i in the form $i = I_{MAX} \sin \omega t$.

[$i = 30 \sin 120\pi t$]

16. An alternating voltage v has a periodic time of 20 ms and a maximum value of 200 V. When time $t = 0$, $v = -75$ volts. Deduce a sinusoidal expression for v and sketch one cycle of the voltage showing important points.

[$v = 200 \sin(100\pi t - 0.384)$]

17 The voltage in an alternating current circuit at any time t seconds is given by $v = 60 \sin 40t$ volts. Find the first time when the voltage is (a) 20 V, and (b) −30 V.

[(a) 8.496 ms; (b) 91.63 ms]

18 The instantaneous value of voltage in an ac circuit at any time t seconds is given by $v = 100 \sin (50\pi t - 0.523)$ V. Find:
(a) the peak-to-peak voltage, the periodic time, the frequency and the phase angle;
(b) the voltage when $t = 0$; (c) the voltage when $t = 8$ ms;
(d) the times in the first cycle when the voltage is 60 V;
(e) the times in the first cycle when the voltage is −40 V; and
(f) the first time when the voltage is a maximum.
Sketch the curve for one cycle showing relevant points.

$$\begin{bmatrix} \text{(a) 200 V, 0.04 s, 25 Hz, } 29°58' \text{ lagging; (b) } -49.95 \text{ V;} \\ \text{(c) 66.96 V; (d) 7.426 ms, 19.23 ms; (e) 25.95 ms, 40.71 ms;} \\ \text{(f) 13.33 ms.} \end{bmatrix}$$

Combination of periodic functions

19 The instantaneous values of two alternating voltages are given by $v_1 = 5 \sin \omega t$ and $v_2 = 8 \sin (\omega t - \frac{\pi}{6})$. By plotting v_1 and v_2 on the same axes, using the same scale, over one cycle, obtain expressions for (a) $v_1 + v_2$ and (b) $v_1 - v_2$.

$$\begin{bmatrix} \text{(a) } v_1 + v_2 = 12.58 \sin (\omega t - 0.325) \text{ V} \\ \text{(b) } v_1 - v_2 = 4.44 \sin (\omega t + 2.02) \text{ V} \end{bmatrix}$$

20 Repeat *Problem 19* by resolution of phasors.

21 Construct a phasor diagram to represent $i_1 + i_2$ where $i_1 = 12 \sin \omega t$ and $i_2 = 15 \sin (\omega t + \pi/3)$. By measurement, or by calculation, find a sinusoidal expression to represent $i_1 + i_2$.

[$23.43 \sin (\omega t + 0.588)$]

22 Determine, either by plotting graphs and adding ordinates at intervals, or by calculation, the following periodic functions in the form $v = V_{MAX} \sin (\omega t \pm \phi)$.
(a) $10 \sin \omega t + 4 \sin (\omega t + \pi/4)$ [(a) $13.14 \sin (\omega t + 0.217)$]
(b) $80 \sin (\omega t + \pi/3) + 50 \sin (\omega t - \pi/6)$ [(b) $94.34 \sin (\omega t + 0.489)$]
(c) $100 \sin \omega t - 70 \sin (\omega t - \pi/3)$ [(c) $88.88 \sin (\omega t + 0.751)$]

Purely resistive ac circuits

23 An alternating voltage of 150 V, 60 Hz is applied across a resistance of 7.5 Ω. Find the current flowing in the circuit and the power dissipated by the resistor.

[20 A; 3 kW]

24 Determine the power developed by a sinusoidal current of peak value 5 A flowing through a resistor of 30 Ω.

[375 W]

25 A sinusoidal emf of maximum value 20 V causes a current of maximum value 2A to flow through a resistance. Find the value of the resistance and the power developed.

[10 Ω; 20 W]

26 A voltage of peak value 300 V is applied across a resistor and a peak current of 50 mA flows. Find the power and the energy dissipated by the resistor in a time of 3 min.

[7.5 W; 1.35 kJ]

Rectifiers

27 Describe, with appropriate circuit diagrams, the principle of operation of (a) a half-wave; and (b) a full-wave rectifier. Explain a simple method of smoothing the output waveform.

7 Single phase AC circuits

A. MAIN POINTS CONCERNED WITH SINGLE PHASE AC CIRCUITS

1. In a purely resistive ac circuit, the current I_R and applied voltage V_R are in phase. See *Fig 1(a)*.
2. In a purely inductive ac circuit, the current I_L **lags** the applied voltage V_L by 90° (i.e. $\pi/2$ rads). See *Fig 1(b)*.
3. In a purely capacitive ac circuit, the current I_C **leads** the applied voltage V_C by 90° (i.e. $\pi/2$ rads). See *Fig 1(c)*.
4. In a purely inductive circuit the opposition to the flow of alternating current is called the **inductive reactance**, X_L.

$$X_L = \frac{V_L}{I_L} = 2\pi f L \text{ ohms}$$

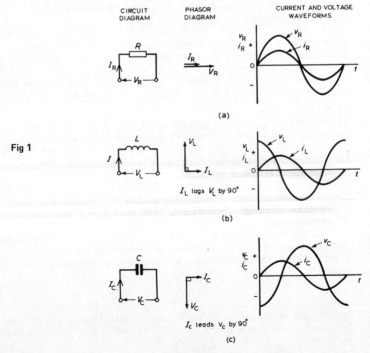

Fig 1

where f is the supply frequency in hertz and L is the inductance in henry's. X_L is proportional to f as shown in *Fig 2(a)*.

5 In a purely capacitive circuit the opposition to the flow of alternating current is called the **capacitive reactance**, X_C.

$$X_C = \frac{V_C}{I_C} = \frac{1}{2\pi f C} \text{ ohms}$$

where C is the capacitance in farads. X_C varies with f as shown in *Fig 2(b)*.

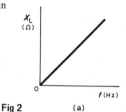
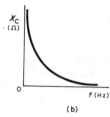

Fig 2   (a)   (b)

6 In an ac series circuit containing inductance L and resistance R, the applied voltage V is the phasor sum of V_R and V_L (see *Fig 3(a)*) and thus the current I lags the applied voltage V by an angle lying between 0° and 90° (depending on the values of V_R and V_L), shown as angle ϕ. In any ac series circuit the current is common to each component and is thus taken as the reference phasor.

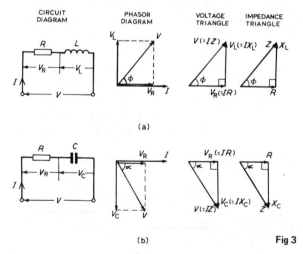

Fig 3

7 In an ac series circuit containing capacitance C and resistance R, the applied voltage V is the phasor sum of V_R and V_C (see *Fig 3(b)*) and thus the current I leads the applied voltage V by an angle lying between 0° and 90° (depending on the values of V_R and V_C), shown as angle α.

8 In an ac circuit, the ratio $\frac{\text{applied voltage } V}{\text{current } I}$ is called the **impedance** Z,

i.e. $Z = \frac{V}{I} \; \Omega$

9 From the phasor diagrams of *Fig 3*, the **'voltage triangles'** are derived.
   (a) For the R–L circuit: $V = \sqrt{(V_R^2 + V_L^2)}$ (by Pythagoras' theorem)
   and $\tan \phi = \dfrac{V_L}{V_R}$ (by trigonometric ratios)

   (b) For the R–C circuit:

   $$V = \sqrt{(V_R^2 + V_C^2)}$$

   and $\tan \alpha = \dfrac{V_C}{V_R}$

10 If each side of the voltage triangles in *Fig 3* is divided by current I then the
   'impedance triangles' are derived.
   (a) For the R–L circuit: $Z = \sqrt{(R^2 + X_L^2)}$
   $\tan \phi = \dfrac{X_L}{R}$, $\sin \phi = \dfrac{X_L}{Z}$ and $\cos \phi = \dfrac{R}{Z}$
   (b) For the R–C circuit: $Z = \sqrt{(R^2 + X_C^2)}$
   $\tan \alpha = \dfrac{X_C}{R}$, and $\sin \alpha = \dfrac{X_C}{Z}$ and $\cos \alpha = \dfrac{R}{Z}$.

11 In an ac series circuit containing resistance R, inductance L and capacitance C,
   the applied voltage V is the phasor sum of V_R, V_L and V_C (see *Fig 4*). V_L and V_C
   are anti-phase and there are three phasor diagrams possible—each depending on
   the relative values of V_L and V_C.

12 When $X_L > X_C$ (see *Fig 4(b)*):
$$Z = \sqrt{[R^2 + (X_L - X_C)^2]}$$
and $\tan \phi = \dfrac{(X_L - X_C)}{R}$

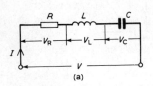

13 When $X_C > X_L$ (see *Fig 4(c)*):
$$Z = \sqrt{[R^2 + (X_C - X_L)^2]}$$
and $\tan \alpha = \dfrac{(X_C - X_L)}{R}$

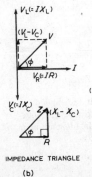

IMPEDANCE TRIANGLE

(b)

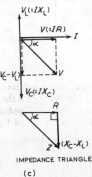

IMPEDANCE TRIANGLE

(c)

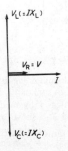

(d)   **Fig 4**

14 When $X_L = X_C$ (see Fig 4(d)), the applied voltage V and the current I are in phase. This effect is called **series resonance**. At resonance:

(i) $V_L = V_C$
(ii) $Z = R$ (i.e. the minimum circuit impedance possible in an L–C–R circuit).
(iii) $I = V/R$ (i.e. the maximum current possible in an L–C–R circuit).
(iv) Since $X_L = X_C$, then $2\pi f_0 L = \dfrac{1}{2\pi f_0 C}$,

from which $f_0 = \dfrac{1}{2\pi\sqrt{(LC)}}$ Hz, where f_0 is the resonant frequency.

15 At resonance, if R is small compared with X_L and X_C, it is possible for V_L and V_C to have voltages many times greater than the supply voltage V (see Fig 4(d)).

Voltage magnification at resonance = $\dfrac{\text{voltage across } L \text{ (or } C)}{\text{supply voltage } V}$

(This magnification is termed the **Q factor**.)

16 (a) For a purely resistive ac circuit, the average power dissipated, P, is given by:

$P = VI = I^2 R = \dfrac{V^2}{R}$ watts (V and I being rms values). See Fig 5(a).

(b) For a purely inductive ac circuit, the average power is zero. See Fig 5(b).
(c) For a purely capacitive ac circuit, the average power is zero. See Fig 5(c).

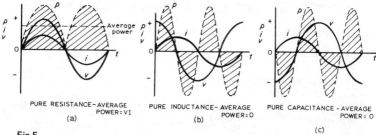

Fig 5

In Figs 5(a)–(c), the value of power at any instant is given by the product of the voltage and current at that instant, i.e. the instantaneous power, $p = vi$, as shown by the broken lines.

17 Fig 6 shows current and voltage waveforms for an R–L circuit where the current lags the voltage by angle ϕ. The waveform for power (where $p = vi$) is shown by the broken line, and its shape, and hence average power, depends on the value of angle ϕ.

For an R–L, R–C or L–C–R series ac circuit, the average power P is given by:

$P = VI \cos\phi$ **watts**

or $P = I^2 R$ **watts** (V and I being rms values).

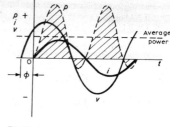

Fig 6

18 *Fig 7(a)* shows a phasor diagram in which the current I lags the applied voltage V by angle ϕ. The horizontal component of V is $V\cos\phi$ and the vertical component of V is $V\sin\phi$. If each of the voltage phasors is multiplied by I, *Fig 7(b)* is obtained and is known as the **power triangle**.

Apparent power, $S = VI$ voltamperes (VA)
True or active power, $P = VI\cos\phi$ watts (W)
Reactive power, $Q = VI\sin\phi$ reactive voltamperes (VAr)

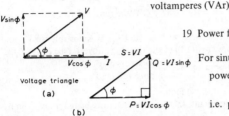

Fig 7

19 Power factor = $\dfrac{\text{True power } P}{\text{Apparent power } S}$

For sinusoidal voltages and currents,

power factor = $\dfrac{P}{S} = \dfrac{VI\cos\phi}{VI}$

i.e. p.f. = $\cos\phi = \dfrac{R}{Z}$ (from *Fig 3*)

(The relationships stated in paras. 18 and 19 are also true when current I leads voltage V.)

B. WORKED PROBLEMS ON SINGLE PHASE AC CIRCUITS

(a) AC CIRCUITS CONTAINING PURE INDUCTANCE AND PURE CAPACITANCE

Problem 1 (a) Calculate the reactance of a coil of inductance 0.32 H when it is connected to a 50 Hz supply.
(b) A coil has a reactance of 124 Ω in a circuit with a supply of frequency 5 kHz. Determine the inductance of the coil.

(a) Inductive reactance, $X_L = 2\pi f L = 2\pi(50)(0.32) = $ **100.5 Ω**

(b) Since $X_L = 2\pi f L$, inductance $L = \dfrac{X_L}{2\pi f} = \dfrac{124}{2\pi(5000)}$ H = **3.95 mH**

Problem 2 A coil has an inductance of 40 mH and negligible resistance. Calculate its inductive reactance and the resulting current if connected to (a) a 240 V, 50 Hz supply, and (b) a 100 V, 1 kHz supply.

(a) Inductive reactance, $X_L = 2\pi f L = 2\pi(50)(40 \times 10^{-3}) = 12.57$ Ω

Current, $I = \dfrac{V}{X_L} = \dfrac{240}{12.57} = $ **19.09 A.**

(b) Inductive reactance, $X_L = 2\pi(1000)(40 \times 10^{-3}) = 251.3$ Ω

Current, $I = \dfrac{V}{X_L} = \dfrac{100}{251.3} = $ **0.398 A.**

Problem 3 Determine the capacitive reactance of a capacitor of 10 μF when connected to a circuit of frequency (a) 50 Hz; (b) 20 kHz.

(a) Capacitive reactance $X_C = \dfrac{1}{2\pi f C} = \dfrac{1}{2\pi(50)(10 \times 10^{-6})} = \dfrac{10^6}{2\pi(50)(10)}$

$= 318.3 \; \Omega$

(b) $X_C = \dfrac{1}{2\pi f C} = \dfrac{1}{2\pi(20 \times 10^3)(10 \times 10^{-6})} = \dfrac{10^6}{2\pi(20 \times 10^3)(10)} = 0.796 \; \Omega$

Hence as the frequency is increased from 50 Hz to 20 kHz, X_C decreases from 318.3 Ω to 0.796 Ω (see *Fig 2(b)*).

Problem 4 A capacitor has a reactance of 40 Ω when operated on a 50 Hz supply. Determine the value of its capacitance.

Since $X_C = \dfrac{1}{2\pi f C}$, capacitance $C = \dfrac{1}{2\pi f X_C} = \dfrac{1}{2\pi(50)(40)}$ F

$= \dfrac{10^6}{2\pi(50)(40)} \; \mu F = 79.58 \; \mu F.$

Problem 5 Calculate the current taken by a 23 μF capacitor when connected to a 240 V, 50 Hz supply.

Current $I = \dfrac{V}{X_C} = \dfrac{V}{\dfrac{1}{2\pi f C}} = 2\pi f C V = 2\pi(50)(23 \times 10^{-6})(240)$

$= 1.73 \; A$

Further problems on ac circuits containing pure inductance and pure capacitance may be found in section C(c), problems 1 to 10, page 114.

(b) *R–L AC CIRCUITS*

Problem 6 In a series R–L circuit the pd across the resistance R is 12 V and the pd across the inductance L is 5 V. Find the supply voltage and the phase angle between current and voltage.

From the voltage triangle of *Fig 3(a)*,
Supply voltage $V = \sqrt{(12^2 + 5^2)}$ i.e. **V = 13 V**
(Note that in ac circuits, the supply voltage is **not** the arithmetic sum of the pd's across components. It is, in fact, the **phasor sum**.)
$\tan \phi = \dfrac{V_L}{V_R} = \dfrac{5}{12}$, from which $\phi = \arctan \dfrac{5}{12} =$ **22° 37′ lagging**.
('Lagging' infers that the current is 'behind' the voltage, since phasors revolve anticlockwise.)

Problem 7 A coil has a resistance of 12 Ω and an inductance of 15.9 mH. Calculate (a) the reactance, (b) the impedance, and (c) the current taken from a 240 V, 50 Hz supply. Determine also the phase angle between the supply voltage and current.

$R = 12\ \Omega;\ L = 15.9\ \text{mH} = 15.9 \times 10^{-3}\ \text{H};\ f = 50\ \text{Hz};\ V = 240\ \text{V}$

(a) Inductive reactance, $X_L = 2\pi f L = 2\pi(50)(15.9 \times 10^{-3}) = 5\ \Omega$

(b) Impedance, $Z = \sqrt{(R^2 + X_L^2)} = \sqrt{(12^2 + 5^2)} \qquad = 13\ \Omega$

(c) Current, $I = \dfrac{V}{Z} = \dfrac{240}{13} = \mathbf{18.5\ A}$

The circuit and phasor diagrams and the voltage and impedance triangles are as shown in *Fig 3(a)*.

Since $\tan \phi = \dfrac{X_L}{R}$, $\phi = \arctan \dfrac{X_L}{R} = \arctan \dfrac{5}{12} = \mathbf{22°\ 37'\ lagging}$

Problem 8 A coil takes a current of 2 A from a 12 V dc supply. When connected to a 240 V, 50 Hz ac supply the current is 20 A. Calculate the resistance, impedance, inductive reactance and inductance of the coil.

Resistance $R = \dfrac{\text{dc voltage}}{\text{dc current}} = \dfrac{12}{2} = 6\ \Omega$.

Impedance $Z = \dfrac{\text{ac voltage}}{\text{ac current}} = \dfrac{240}{20} = 12\ \Omega$.

Since $Z = \sqrt{(R^2 + X_L^2)}$, inductive reactance, $X_L = \sqrt{(Z^2 - R^2)}$
$\qquad\qquad\qquad\qquad\qquad\qquad\qquad\qquad = \sqrt{(12^2 - 6^2)} = 10.39\ \Omega$

Since $X_L = 2\pi f L$, inductance $L = \dfrac{X_L}{2\pi f} = \dfrac{10.39}{2\pi(50)} = \mathbf{33.1\ mH}$

This problem indicates a simple method for finding the inductance of a coil, i.e. firstly to measure the current when the coil is connected to a dc supply of known voltage, and then to repeat the process with an ac supply.

Problem 9 A coil of inductance 318.3 mH and negligible resistance is connected in series with a 200 Ω resistor to a 240 V, 50 Hz supply. Calculate (a) the inductive reactance of the coil; (b) the impedance of the circuit; (c) the current in the circuit; (d) the pd across each component, and (e) the circuit phase angle.

$L = 318.3\ \text{mH} = 0.3183\ \text{H};\ R = 200\ \Omega;\ V = 240\ \text{V};\ f = 50\ \text{Hz}.$

The circuit diagram is as shown in *Fig 3(a)*.

(a) Inductive reactance $X_L = 2\pi f L = 2\pi(50)(0.3183) = \mathbf{100\ \Omega}$

(b) Impedance $Z = \sqrt{(R^2 + X_L^2)} = \sqrt{[(200)^2 + (100)^2]} = \mathbf{223.6\ \Omega}$

(c) Current $I = \dfrac{V}{Z} = \dfrac{240}{223.6} = \mathbf{1.073\ A}$

(d) The pd across the coil, $V_L = IX_L = 1.073 \times 100 = \textbf{107.3 V}$
The pd across the resistor, $V_R = IR = 1.073 \times 200 = \textbf{214.6 V}$

[Check: $\sqrt{(V_R^2+V_L^2)} = \sqrt{[(214.6)^2+(107.3)^2]} = 240$ V, the supply voltage]

(e) From the impedance triangle, angle $\phi = \arctan \dfrac{X_L}{R} = \arctan \dfrac{100}{200}$

Hence phase angle $\phi = \textbf{26° 34' lagging}$.

Problem 10 A coil consists of a resistance of 100 Ω and an inductance of 200 mH. If an alternating voltage, v, given by $v = 200 \sin 500\,t$ volts is applied across the coil, calculate (a) the circuit impedance, (b) the current flowing, (c) the pd across the resistance, (d) the pd across the inductance and (e) the phase angle between voltage and current.

Since $v = 200 \sin 500\,t$ volts then $V_{MAX} = 200$ V and $\omega = 2\pi f = 500$ rad/s.
Hence rms voltage $V = 0.707 \times 200 = 141.4$ V.
Inductive reactance, $X_L = 2\pi f L = \omega L = 500 \times 200 \times 10^{-3} = 100\,\Omega$

(a) Impedance $Z = \sqrt{(R^2+X_L^2)} = \sqrt{(100^2+100^2)} = \textbf{141.4 Ω}$

(b) Current $I = \dfrac{V}{Z} = \dfrac{141.4}{141.4} = \textbf{1 A}$

(c) pd across the resistance $V_R = IR = 1 \times 100 = \textbf{100 V}$

(d) pd across the inductance $V_L = IX_L = 1 \times 100 = \textbf{100 V}$

(e) Phase angle between voltage and current, $\phi = \dfrac{X_L}{R} = \arctan \dfrac{100}{100}$

Hence $\phi = \textbf{45°}$ or $\dfrac{\pi}{4}$ **rads**.

Further problems on R–L ac circuits may be found in section C(c), Problems 11 to 16, page 115.

(c) *R–C AC CIRCUITS*

Problem 11 A resistor of 25 Ω is connected in series with a capacitor of 45 μF. Calculate (a) the impedance; and (b) the current taken from a 240 V, 50 Hz supply. Find also the phase angle between the supply voltage and the current.

$R = 25\,\Omega$; $C = 45\,\mu\text{F} = 45 \times 10^{-6}$ F; $V = 240$ V; $f = 50$ Hz.
The circuit diagram is as shown in *Fig 3(b)*.

Capacitive reactance, $X_C = \dfrac{1}{2\pi f C} = \dfrac{1}{2\pi(50)(45 \times 10^{-6})} = 70.74\,\Omega$

(a) Impedance $Z = \sqrt{(R^2+X_C^2)} = \sqrt{[(25)^2+(70.74)^2]} = \textbf{75.03 Ω}$

(b) Current $I = \dfrac{V}{Z} = \dfrac{240}{75.03} = \textbf{3.20 A}$.

Phase angle between the supply voltage and current, $\alpha = \arctan \dfrac{X_C}{R}$

Hence $\alpha = \arctan \dfrac{70.74}{25} = \mathbf{70°\ 32'\ leading}$.

('Leading' infers that the current is 'ahead' of the voltage, since phasors revolve anticlockwise.)

Problem 12 A capacitor C is connected in series with a 40 Ω resistor across a supply of frequency 60 Hz. A current of 3 A flows and the circuit impedance is 50 Ω. Calculate:
(a) the value of capacitance, C; (b) the supply voltage; (c) the phase angle between the supply voltage and current; (d) the pd across the resistor, and (e) the pd across the capacitor. Draw the phasor diagram.

(a) Impedance $Z = \sqrt{(R^2 + X_C^2)}$

Hence $X_C = \sqrt{(Z^2 - R^2)}$
$= \sqrt{(50^2 - 40^2)} = 30\ \Omega$

$X_C = \dfrac{1}{2\pi f C}$

Hence $C = \dfrac{1}{2\pi f X_C} = \dfrac{1}{2\pi(60)30}\ \text{F} = \mathbf{88.42\ \mu F}$

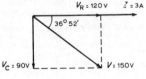

Phasor diagram

Fig 8

(b) Since $Z = \dfrac{V}{I}$ then $V = IZ = (3)(50) = \mathbf{150\ V}$

(c) Phase angle, $\alpha = \arctan \dfrac{X_C}{R} = \arctan \dfrac{30}{40} = \mathbf{36°\ 52'\ leading}$

(d) pd across resistor, $V_R = IR = (3)(40) = \mathbf{120\ V}$

(e) pd across capacitor, $V_C = IX_C = (3)(30) = \mathbf{90\ V}$

The phasor diagram is shown in *Fig 8*.

Further problems on R–C ac circuits may be found in section C(c), Problems 17 to 21, page 115.

(d) L–C–R AC CIRCUITS AND SERIES RESONANCE

Problem 13 A coil of resistance 5 Ω and inductance 120 mH in series with a 100 μF capacitor, is connected to a 300 V, 50 Hz supply. Calculate (a) the current flowing; (b) the phase difference between the supply voltage and current; (c) the voltage across the coil and (d) the voltage across the capacitor.

The circuit diagram is shown in *Fig 9*.

$X_L = 2\pi f L = 2\pi(50)(120 \times 10^{-3}) = 37.70\ \Omega$

$X_C = \dfrac{1}{2\pi f C} = \dfrac{1}{2\pi(50)(100 \times 10^{-6})} = 31.83\ \Omega$

Since X_L is greater than X_C the circuit is inductive (see phasor diagram in *Fig 4(b)*).

$X_L - X_C = 37.70 - 31.83 = 5.87 \ \Omega$

Impedance $Z = \sqrt{[R^2 + (X_L - X_C)^2]}$

$\qquad = \sqrt{[(5)^2 + (5.87)^2]} = 7.71 \ \Omega$

(a) Current $I = \dfrac{V}{Z} = \dfrac{300}{7.71} = \mathbf{38.91 \ A}$

(b) Phase angle ϕ

$\qquad = \arctan \left(\dfrac{X_L - X_C}{R} \right)$

$\qquad = \arctan \left(\dfrac{5.87}{5} \right) = \mathbf{49° \ 35'}$

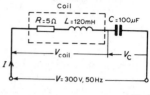

Fig 9

(c) Impedance of coil $Z_{COIL} = \sqrt{(R^2 + X_L^2)} = \sqrt{[(5)^2 + (37.7)^2]} = 38.03 \ \Omega$
Voltage across coil $V_{COIL} = I Z_{COIL} = (38.91)(38.03) = \mathbf{1480 \ V}$

(d) Voltage across capacitor $V_C = I X_C = (38.91)(31.83) = \mathbf{1239 \ V}$

Problem 14 A coil having a resistance of 10 Ω and an inductance of 125 mH is connected in series with a 60 μF capacitor across a 120 V supply. At what frequency does resonance occur? Find the current flowing at the resonant frequency.

Resonant frequency, $f_0 = \dfrac{1}{2\pi \sqrt{(LC)}}$ Hz $= \dfrac{1}{2\pi \sqrt{\left[\left(\dfrac{125}{10^3}\right) \left(\dfrac{60}{10^6}\right) \right]}}$ Hz

$\qquad = \dfrac{1}{2\pi \sqrt{\left(\dfrac{125 \times 6}{10^8}\right)}} = \dfrac{1}{\dfrac{2\pi \sqrt{(125)(6)}}{10^4}}$

$\qquad = \dfrac{10^4}{2\pi \sqrt{(125)(6)}} = \mathbf{58.12 \ Hz}$

At resonance, $X_L = X_C$ and impedance $Z = R$

Hence current, $I = \dfrac{V}{R} = \dfrac{120}{10} = \mathbf{12 \ A}$

Problem 15 The current at resonance in a series $L-C-R$ circuit is 100 μA. If the applied voltage is 2 mV at a frequency of 200 kHz, and the circuit inductance is 50 μH, find (a) the circuit resistance, and (b) the circuit capacitance.

(a) $I = 100 \ \mu A = 100 \times 10^{-6}$ A; $V = 2$ mV $= 2 \times 10^{-3}$ V
At resonance, impedance $Z =$ resistance R

Hence $R = \dfrac{V}{I} = \dfrac{2 \times 10^{-3}}{100 \times 10^{-6}} = \dfrac{2 \times 10^6}{100 \times 10^3} = \mathbf{20 \ \Omega}$

(b) At resonance $X_L = X_C$

i.e. $2\pi fL = \dfrac{1}{2\pi fC}$

Hence capacitance $C = \dfrac{1}{(2\pi f)^2 L} = \dfrac{1}{(2\pi \times 200 \times 10^3)^2 (50 \times 10^{-6})}$ F

$= \dfrac{(10^6)(10^6)}{(4\pi)^2(10^{10})(50)}$ μF = **0.0127 μF or 12.7 nF**

Problem 16 A coil of inductance 80 mH and negligible resistance is connected in series with a capacitance of 0.25 μF and a resistor of resistance 12.5 Ω across a 100 V, variable frequency supply. Determine (a) the resonant frequency, and (b) the current at resonance. How many times greater than the supply voltage is the voltage across the reactances at resonance?

(a) Resonant frequency $f_0 = \dfrac{1}{2\pi\sqrt{\left[\left(\dfrac{80}{10^3}\right)\left(\dfrac{0.25}{10^6}\right)\right]}} = \dfrac{1}{2\pi\sqrt{\dfrac{(8)(0.25)}{10^8}}} = \dfrac{10^4}{2\pi\sqrt{2}}$

= **1125.4 Hz = 1.1254 kHz**

(b) Current at resonance $I = \dfrac{V}{R} = \dfrac{100}{12.5} =$ **8 A**

Voltage across inductance, at resonance, $V_L = IX_L = (I)(2\pi fL)$
$= (8)(2\pi)(1125.4)(80 \times 10^{-3})$
$=$ **4525.5 V**

$$\left[\text{Also, voltage across capacitor, } V_C = IX_C = \dfrac{I}{2\pi fC} = \dfrac{8}{2\pi(1125.4)(0.25 \times 10^{-6})} = 4525.5 \text{ V} \right]$$

Voltage magnification at resonance $= \dfrac{V_L}{V}$ or $\dfrac{V_C}{V} = \dfrac{4525.5}{100} =$ **45.255**

i.e. at resonance, the voltages across the reactances are 45.255 times greater than the supply voltage. Hence Q factor of circuit is 45.255.

Further problems on L–C–R ac circuits and resonance may be found in section C(c), Problems 22 to 27, page 116.

(e) POWER IN AC CIRCUITS

Problem 17 An instantaneous current, $i = 250 \sin \omega t$ mA flows through a pure resistance of 5 kΩ. Find the power dissipated in the resistor.

Power dissipated, $P = I^2 R$ where I is the rms value of current.
If $i = 250 \sin \omega t$ mA, then $I_{MAX} = 0.250$ A and rms current, $I = (0.707 \times 0.250)$ A
Hence power $P = (0.707 \times 0.250)^2 (5000) =$ **156.2 watts**

Problem 18 A series circuit of resistance $60\,\Omega$ and inductance $75\,\text{mH}$ is connected to a 110 V, 60 Hz supply. Calculate the power dissipated.

Inductive reactance, $X_L = 2\pi f L = 2\pi(60)(75 \times 10^{-3}) = 28.27\,\Omega$
Impedance, $Z = \sqrt{(R^2 + X_L^2)} = \sqrt{[(60)^2 + (28.27)^2]} = 66.33\,\Omega$
Current, $I = \dfrac{V}{Z} = \dfrac{110}{66.33} = 1.658\,\text{A}$

To calculate power dissipation in an ac circuit two formulae may be used:

(i) $P = I^2 R = (1.658)^2 (60) = $ **165 W**

or (ii) $P = VI \cos\phi$ where $\cos\phi = \dfrac{R}{Z} = \dfrac{60}{66.33} = 0.9046$

Hence $P = (110)(1.658)(0.9046) = $ **165 W**

Problem 19 A pure inductance is connected to a 150 V, 50 Hz supply, and the apparent power of the circuit is 300 VA. Find the value of the inductance.

Apparent power $S = VI$
Hence current $I = \dfrac{S}{V} = \dfrac{300}{150} = 2\,\text{A}$
Inductive reactance $X_L = \dfrac{V}{I} = \dfrac{150}{2} = 75\,\Omega$
Since $X_L = 2\pi f L$, inductance $L = \dfrac{X_L}{2\pi f} = \dfrac{75}{2\pi(50)} = $ **0.239 H**

Problem 20 A transformer has a rated output of 200 kVA at a power factor of 0.8. Determine the rated power output and the corresponding reactive power.

$VI = 200\,\text{kVA} = 200 \times 10^3$; p.f. $= 0.8 = \cos\phi$.
Power output, $P = VI \cos\phi = (200 \times 10^3)(0.8) = $ **160 kW**
Reactive power, $Q = VI \sin\phi$
If $\cos\phi = 0.8$, then $\phi = \arccos 0.8 = 36°\,52'$
Hence $\sin\phi = \sin 36°\,52' = 0.6$
Hence reactive power, $Q = (200 \times 10^3)(0.6) = $ **120 kVAr**

Problem 21 A load takes 90 kW at a power factor of 0.5 lagging. Calculate the apparent power and the reactive power.

True power $P = 90$ kW $= VI \cos \phi$ Power factor $= 0.5 = \cos \phi$
Apparent power $S = VI = \dfrac{90 \text{ kW}}{0.5} = \mathbf{180}$ **kVA**

Angle $\phi = \arccos 0.5 = 60°$. Hence $\sin \phi = \sin 60° = 0.866$
Hence reactive power, $Q = VI \sin \phi = 180 \times 10^3 \times 0.866 = \mathbf{156}$ **kVAr**

Problem 22 The power taken by an inductive circuit when connected to a 120 V, 50 Hz supply is 400 W and the current is 8 A. Calculate (a) the resistance, (b) the impedance; (c) the reactance; (d) the power factor, and (e) the phase angle between voltage and current.

(a) Power $P = I^2 R$. Hence $R = \dfrac{P}{I^2} = \dfrac{400}{(8)^2} = \mathbf{6.25 \ \Omega}$

(b) Impedance $Z = \dfrac{V}{I} = \dfrac{120}{8} = \mathbf{15 \ \Omega}$

(c) Since $Z = \sqrt{(R^2 + X_L^2)}$, then $X_L = \sqrt{(Z^2 - R^2)} = \sqrt{[(15)^2 - (6.25)^2]} = \mathbf{13.64 \ \Omega}$

(d) Power factor $= \dfrac{\text{true power}}{\text{apparent power}} = \dfrac{VI \cos \phi}{VI} = \dfrac{400}{(120)(8)} = \mathbf{0.4167}$

(e) p.f. $= \cos \phi = 0.4167$. Hence phase angle $\phi = \arccos 0.4167 = \mathbf{65° \ 22'}$ **lagging**

Problem 23 A circuit consisting of a resistor in series with a capacitor takes 100 watts at a power factor of 0.5 from a 100 V, 60 Hz supply. Find (a) the current flowing; (b) the phase angle; (c) the resistance; (d) the impedance; and (e) the capacitance.

(a) Power factor $= \dfrac{\text{true power}}{\text{apparent power}}$

i.e. $0.5 = \dfrac{100}{100 \, I}$. Hence $I = \dfrac{100}{(0.5)(100)} = \mathbf{2 \ A}$

(b) Power factor $= 0.5 = \cos \phi$. Hence phase angle $\phi = \arccos 0.5 = \mathbf{60°}$ **leading**

(c) Power $P = I^2 R$. Hence resistance $R = \dfrac{P}{I^2} = \dfrac{100}{(2)^2} = \mathbf{25 \ \Omega}$

(d) Impedance $Z = \dfrac{V}{I} = \dfrac{100}{2} = \mathbf{50 \ \Omega}$

(e) Capacitive reactance, $X_C = \sqrt{(Z^2 - R^2)} = \sqrt{(50^2 - 25^2)} = 43.30 \ \Omega$

$X_C = \dfrac{1}{2\pi f C}$. Hence capacitance $C = \dfrac{1}{2\pi f X_C} = \dfrac{1}{2\pi (60)(43.30)}$ F $= \mathbf{61.26 \ \mu F}$

Further problems on power in ac circuits may be found in section C(c), Problems 28 to 39, page 116.

C. FURTHER PROBLEMS ON SINGLE PHASE AC CIRCUITS

(a) SHORT ANSWER PROBLEMS

1. Complete the following statements:
   (a) In a purely resistive ac circuit the current is .... with the voltage.
   (b) In a purely inductive ac circuit the current .... the voltage by .... degrees.
   (c) In a purely capacitive ac circuit the current .... the voltage by .... degrees.
2. Draw phasor diagrams to represent (a) a purely resistive ac circuit, (b) a purely inductive ac circuit, and (c) a purely capacitive ac circuit.
3. What is inductive reactance? State the symbol and formula for determining inductive reactance.
4. What is capacitive reactance? State the symbol and formula for determining capacitive reactance.
5. What does 'impedance' mean when referring to an ac circuit?
6. Draw an impedance triangle for an R–L circuit. Derive from the triangle an expression for (a) impedance, and (b) phase angle.
7. State two formulae which may be used to calculate power in an ac circuit.
8. What is series resonance?
9. Derive a formula for resonant frequency f_0 in terms of L and C.
10. Define 'power factor'.
11. Define (a) apparent power; (b) reactive power.
12. Show graphically that for a purely inductive or purely capacitive ac circuit the average power is zero.

(b) MULTI-CHOICE PROBLEMS (Answers on page 152)

1. An inductance of 10 mH connected across a 100 V, 50 Hz supply has an inductive reactance of (a) 10π Ω; (b) 1000π Ω; (c) π Ω; (d) π H.
2. When the frequency of an ac circuit containing resistance and inductance is increased, the current (a) decreases; (b) increases; (c) stays the same.
3. In *Problem 2*, the phase angle of the circuit (a) decreases; (b) increases; (c) stays the same.
4. A capacitor of 1 μF is connected to a 50 Hz supply. The capacitive reactance is
   (a) 50 MΩ; (b) $\dfrac{10}{\pi}$ kΩ; (c) $\dfrac{\pi}{10^4}$ Ω; (d) $\dfrac{10}{\pi}$
5. In a series ac circuit the voltage across a pure inductance is 12 V and the voltage across a pure resistance is 5 V. The supply voltage is
   (a) 13 V; (b) 17 V; (c) 7 V; (d) 2.4 V.
6. Inductive reactance results in a current that
   (a) leads the voltage by 90°;
   (b) is in phase with the voltage;
   (c) leads the voltage by π rads;
   (d) lags the voltage by $\pi/2$ rads.

7   State which of the following statements is false.
   (a) Impedance is at a minimum at resonance in an ac circuit.
   (b) The product of rms current and voltage gives the apparent power in an ac circuit.
   (c) Current is at a maximum at resonance in an ac circuit.
   (d) $\dfrac{\text{Apparent power}}{\text{True power}}$ gives power factor.

8   The impedance of a coil, which has a resistance of X ohms and an inductance of Y henry's, connected across a supply of frequency K Hz is
   (a) $2\pi KY$; (b) $X+Y$; (c) $\sqrt{(X^2+Y^2)}$; (d) $\sqrt{\{X^2+(2\pi KY)^2\}}$.

9   In *Problem 8*, the phase angle between the current and the applied voltage is given by:
   (a) $\arctan \dfrac{Y}{X}$; (b) $\arctan\left(\dfrac{2\pi KY}{X}\right)$; (c) $\arctan\left(\dfrac{X}{2\pi KY}\right)$; (d) $\tan\left(\dfrac{2\pi KY}{X}\right)$.

10 When a capacitor is connected to an ac supply the current
   (a) leads the voltage by 180°; (b) is in phase with the voltage; (c) leads the voltage by $\pi/2$ rads; (d) lags the voltage by 90°.

11 When the frequency of an ac circuit containing resistance and capacitance is decreased the current (a) increases; (b) decreases; (c) stays the same.

12 In an R–L–C ac circuit a current of 5 A flows when the supply voltage is 100 V. The phase angle between current and voltage is 60° lagging. Which of the following statements is false?
   (a) The circuit is effectively inductive.
   (b) The apparent power is 500 VA
   (c) The equivalent circuit reactance is 20 Ω.
   (d) The true power is 250 W.

(c) CONVENTIONAL PROBLEMS

ac circuits containing pure inductance and pure capacitance

1   Calculate the reactance of a coil of inductance 0.2 H when it is connected to (a) a 50 Hz; (b) a 600 Hz and (c) a 40 kHz supply.
   [(a) 62.83 Ω; (b) 754 Ω; (c) 50.27 kΩ]

2   A coil has a reactance of 120 Ω in a circuit with a supply frequency of 4 kHz. Calculate the inductance of the coil.
   [4.77 mH]

3   A supply of 240 V, 50 Hz is connected across a pure inductance and the resulting current is 1.2 A. Calculate the inductance of the coil.
   [0.637 H]

4   An emf of 200 V at a frequency of 2 kHz is applied to a coil of pure inductance 50 mH. Determine (a) the reactance of the coil, and (b) the current flowing in the coil.
   [(a) 628 Ω; (b) 0.318 A]

5   A 120 mH inductor has a 50 mA, 1 kHz alternating current flowing through it. Find the pd across the inductor.
   [37.7 V]

6   Calculate the capacitive reactance of a capacitor of 20 μF when connected to an ac circuit of frequency (a) 20 Hz; (b) 500 Hz; (c) 4 kHz.
   [(a) 397.9 Ω; (b) 15.92 Ω; (c) 1.989 Ω]

7 A capacitor has a reactance of 80 Ω when connected to a 50 Hz supply. Calculate the value of its capacitance.

[39.79 μF]

8 Calculate the current taken by a 10 μF capacitor when connected to a 200 V, 100 Hz supply.

[1.257 A]

9 A capacitor has a capacitive reactance of 400 Ω when connected to a 100 V, 25 Hz supply. Determine its capacitance and the current taken from the supply.

[15.92 μF; 0.25 A]

10 Two similar capacitors are connected in parallel to a 200 V, 1 kHz supply. Find the value of each capacitor if the circuit current is 0.628 A.

[0.25 μF]

R–L ac circuits

11 Determine the impedance of a coil which has a resistance of 12 Ω and a reactance of 16 Ω.

[20 Ω]

12 A coil of inductance 80 mH and resistance 60 Ω is connected to a 200 V, 100 Hz supply. Calculate the circuit impedance and the current taken from the supply. Find also the phase angle between the current and the supply voltage.

[78.27 Ω; 2.555 A; 39° 57′ lagging]

13 An alternating voltage given by $v = 100 \sin 240\, t$ volts is applied across a coil of resistance 32 Ω and inductance 100 mH. Determine (a) the circuit impedance; (b) the current flowing; (c) the pd across the resistance, and (d) the pd across the inductance.

[(a) 40 Ω; (b) 1.77 A; (c) 56.64 V; (d) 42.48 V]

14 A coil takes a current of 5 A from a 20 V dc supply. When connected to a 200 V, 50 Hz ac supply the current is 25 A. Calculate the (a) resistance; (b) impedance and (c) inductance of the coil.

[(a) 4 Ω; (b) 8 Ω; (c) 22.06 mH]

15 A resistor and an inductor of negligible resistance are connected in series to an ac supply. The pd across the resistor is 18 V and the pd across the inductor is 24 V. Calculate the supply voltage and the phase angle between voltage and current.

[30 V, 53° 8′ lagging]

16 A coil of inductance 636.6 mH and negligible resistance is connected in series with a 100 Ω resistor to a 240 V, 50 Hz supply. Calculate (a) the inductive reactance of the coil; (b) the impedance of the circuit, (c) the current in the circuit; (d) the pd across each component, and (e) the circuit phase angle.

[(a) 200 Ω; (b) 223.6 Ω; (c) 1.118 A; (d) 223.6 V, 111.8 V; (e) 63° 26′ lagging]

R–C ac circuits

17 A voltage of 35 V is applied across a *C–R* series circuit. If the voltage across the resistor is 21 V, find the voltage across the capacitor.

[28 V]

18 A resistance of 50 Ω is connected in series with a capacitance of 20 μF. If a supply of 200 V, 100 Hz is connected across the arrangement find (a) the circuit impedance; (b) the current flowing; and (c) the phase angle between voltage and current.

[(a) 93.98 Ω; (b) 2.128 A; (c) 57° 51′ leading]

19. A 24.87 µF capacitor and a 30 Ω resistor are connected in series across a 150 V supply. If the current flowing is 3 A find (a) the frequency of the supply, (b) the pd across the resistor and (c) the pd across the capacitor.

[(a) 160 Hz; (b) 90 V; (c) 120 V]

20. An alternating voltage $v = 250 \sin 800\,t$ volts is applied across a series circuit containing a 30 Ω resistor and 50 µF capacitor. Calculate (a) the circuit impedance; (b) the current flowing; (c) the pd across the resistor; (d) the pd across the capacitor; and (e) the phase angle between voltage and current.

[(a) 39.05 Ω; (b) 4.526 A; (c) 135.8 V; (d) 113.2 V; (e) 39° 48']

21. A 400 Ω resistor is connected in series with a 2358 pF capacitor across a 12 V ac supply. Determine the supply frequency if the current flowing in the circuit is 24 mA.

[225 kHz]

L–C–R ac circuits and series resonance

22. A 40 µF capacitor in series with a coil of resistance 8 Ω and inductance 80 mH is connected to a 200 V, 100 Hz supply. Calculate (a) the circuit impedance; (b) the current flowing; (c) the phase angle between voltage and current; (d) the voltage across the coil; and (e) the voltage across the capacitor.

[(a) 13.18 Ω; (b) 15.17 A; (c) 52° 38'; (d) 772.1 V; (e) 603.6 V]

23. Find the resonant frequency of a series ac circuit consisting of a coil of resistance 10 Ω and inductance 50 mH and capacitance 0.05 µF. Find also the current flowing at resonance if the supply voltage is 100 V.

[3.183 kHz; 10 A]

24. The current at resonance in a series L–C–R circuit is 0.2 mA. If the applied voltage is 250 mV at a frequency of 100 kHz and the circuit capacitance is 0.04 µF, find the circuit resistance and inductance.

[1.25 kΩ; 63.3 µH]

25. A coil of resistance 25 Ω and inductance 100 mH is connected in series with a capacitance of 0.12 µF across a 200 V, variable frequency supply. Calculate (a) the resonant frequency; (b) the current at resonance and (c) the factor by which the voltage across the reactance is greater than the supply voltage.

[(a) 1.453 kHz; (b) 8 A; (c) 36.52]

26. A coil of 0.5 H inductance and 8 Ω resistance is connected in series with a capacitor across a 200 V, 50 Hz supply. If the current is in phase with the supply voltage, determine the value of the capacitor and the pd across its terminals.

[20.26 µF; 3.928 kV]

27. Calculate the inductance which must be connected in series with a 1000 pF capacitor to give a resonant frequency of 400 kHz.

[0.158 mH]

Power in ac circuits

28. A voltage $v = 200 \sin \omega t$ volts is applied across a pure resistance of 1.5 kΩ. Find the power dissipated in the resistor.

[13.33 W]

29. A 50 µF capacitor is connected to a 100 V, 200 Hz supply. Determine the true power and the apparent power.

[0; 628.3 VA]

30. A motor takes a current of 10 A when supplied from a 250 V ac supply. Assuming a power factor of 0.75 lagging find the power consumed. Find also the cost of running the motor for 1 week continuously if 1 kWh of electricity costs 4.20 p.

[1875 W; £13.23]

31 A transformer has a rated output of 100 kVA at a power factor of 0.6. Determine the rated power output and the corresponding reactive power.

[60 kW; 80 kVAr]

32 A substation is supplying 200 kVA and 150 kVAr. Calculate the corresponding power and power factor.

[132 kW; 0.66]

33 A load takes 50 kW at a power factor of 0.8 lagging. Calculate the apparent power and the reactive power.

[62.5 kVA; 37.5 kVAr]

34 A coil of resistance 400 Ω and inductance 0.20 H is connected to a 75 V, 400 Hz supply. Calculate the power dissipated in the coil.

[5.452 W]

35 An 80 Ω resistor and a 6 μF capacitor are connected in series across a 150 V, 200 Hz supply. Calculate (a) the circuit impedance; (b) the current flowing and (c) the power dissipated in the circuit.

[(a) 154.9 Ω; (b) 0.968 A; (c) 75 W]

36 The power taken by a series circuit containing resistance and inductance is 240 W when connected to a 200 V, 50 Hz supply. If the current flowing is 2 A find the values of the resistance and inductance.

[60 Ω; 255 mH]

37 The power taken by a C–R series circuit, when connected to a 105 V, 2.5 kHz supply, is 0.9 kW and the current is 15 A. Calculate (a) the resistance; (b) the impedance; (c) the reactance; (d) the capacitance; (e) the power factor and (f) the phase angle between voltage and current.

[(a) 4 Ω; (b) 7 Ω; (c) 5.745 Ω; (d) 11.08 μF; (e) 0.571; (f) 55° 9' leading]

38 A circuit consisting of a resistor in series with an inductance takes 210 W at a power factor of 0.6 from a 50 V, 100 Hz supply. Find (a) the current flowing; (b) the circuit phase angle; (c) the resistance; (d) the impedance, and (e) the inductance.

[(a) 7 A; (b) 53° 8' lagging; (c) 4.286 Ω; (d) 7.143 Ω; (e) 9.095 mH]

39 A 200 V, 60 Hz supply is applied to a capacitive circuit. The current flowing is 2 A and the power dissipated is 150 W. Calculate the values of the resistance and capacitance.

[37.5 Ω; 28.61 μF]

8 Semiconductor diodes

A MAIN POINTS ASSOCIATED WITH SEMICONDUCTOR DIODES

1 Materials may be classified as **conductors, semiconductors** or **insulators**. The classification depends on the value of resistivity of the material. Good conductors are usually metals and have resistivities in the order of 10^{-7} to 10^{-8} Ωm. Semiconductors have resistivities in the order of 10^{-3} to 3×10^{3} Ωm. The resistivities of insulators are in the order of 10^{4} to 10^{14} Ω. Some typical approximate values at normal room temperatures are:

Conductors:
  Aluminium 2.7×10^{-8} Ωm
  Brass (70 Cu/30 Zn) 8×10^{-8} Ωm
  Copper (pure annealed) 1.7×10^{-8} Ωm
  Steel (mild) 15×10^{-8} Ωm

Semiconductors:
  Silicon 2.3×10^{3} Ωm ⎫ at 27 °C
  Germanium 0.45 Ωm   ⎭

Insulators:
  Glass $\geqslant 10^{10}$ Ωm
  Mica $\geqslant 10^{11}$ Ωm
  PVC $\geqslant 10^{13}$ Ωm
  Rubber (pure) 10^{12} to 10^{14} Ωm

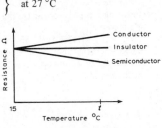

Fig 1

2 In general, over a limited range of temperatures, the resistance of a conductor increases with temperature increase. The resistance of insulators remains approximately constant with variation of temperature. The resistance of semiconductor materials decreases as the temperature increases. For a specimen of each of these materials, having the same resistance (and thus completely different dimensions), at, say, 15 °C, the variation for a small increase in temperature to t °C is as shown in *Fig 1*.

3 The most important semiconductors used in the electronics industry are **silicon** and **germanium**. As the temperature of these materials is raised above room temperature, the resistivity is reduced and ultimately a point is reached where they effectively become conductors. For this reason, silicon should not operate at a working temperature in excess of 150 °C to 200 °C, depending on its purity, and germanium should not operate at a working temperature in excess of 75 °C to 90 °C, depending on its purity. As the temperature of a semiconductor is reduced below normal

room temperature, the resistivity increases until, at very low temperatures, the semiconductor becomes an insulator.

4 Adding extremely small amounts of impurities to pure semiconductors in a controlled manner is called **doping**. Antimony, arsenic and phosphorus are called *n*-type impurities and form an ***n*-type material** when any of these impurities are added to silicon or germanium. The amount of impurity added usually varies from 1 part impurity in 10^5 parts semiconductor material to 1 part impurity to 10^8 parts semiconductor material, depending on the resistivity required. Indium, aluminium and boron are called p-type impurities and form a *p*-**type material** when any of these impurities are added to a semiconductor.

5 In semiconductor materials, there are very few charge carriers per unit volume free to conduct. This is because the 'four electron structure' in the outer shell of the atoms (called valency electrons), form strong covalent bonds with neighbouring atoms, resulting in a tetrahedral structure with the electrons held fairly rigidly in place. A two-dimensional diagram depicting this is shown for germanium in *Fig 2*.

Arsenic, antimony and phosphorus have five valency electrons and when a semiconductor is doped with one of these substances, some impurity atoms are incorporated in the tetrahedral structure. The 'fifth' valency electron is not rigidly

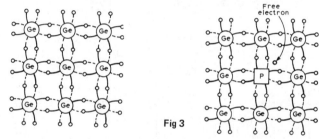

Fig 2 **Fig 3**

bonded and is free to conduct, the impurity atom donating a charge carrier. A two-dimensional diagram depicting this is shown in *Fig 3*, in which a phosphorus atom has replaced one of the germanium atoms. The resulting material is called *n*-type material, and contains free electrons.

Indium, aluminium and boron have three valency electrons and when a semiconductor is doped with one of these substances some of the semiconductor atoms

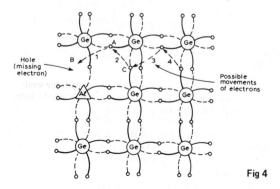

Fig 4

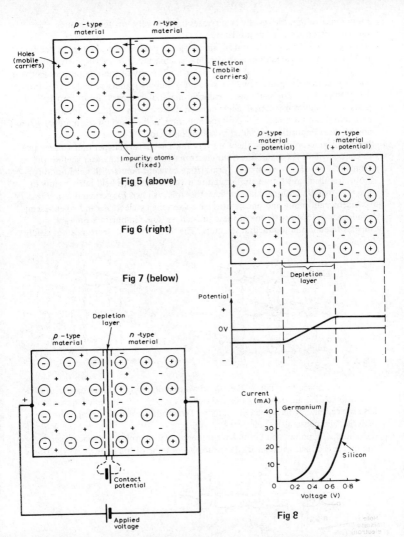

Fig 5 (above)

Fig 6 (right)

Fig 7 (below)

Fig 8

are replaced by impurity atoms. One of the four bonds associated with the semiconductor material is deficient by one electron and this deficiency is called a **hole**.

Holes give rise to conduction when a potential difference exists across the semiconductor material due to movement of electrons from one hole to another, as shown in *Fig 4*. In this figure, an electron moves from A to B, giving the appearance that the hole moves from B to A. Then electron C moves to A, giving the appearance that the hole moves to C, and so on. The resulting material is *p*-type material containing holes.

6   **A p–n junction** is a piece of semiconductor material in which part of the material is *p*-type and part is *n*-type. In order to examine the charge situation, assume that separate blocks of *p*-type and *n*-type materials are pushed together. Also assume that a hole is a positive charge carrier and that an electron is a negative charge carrier.

   At the junction, the donated electrons in the *n*-type material, called **majority carriers**, diffuse into the *p*-type material (diffusion is from an area of high density to an area of lower density) and the acceptor holes in the *p*-type material diffuse into the *n*-type material as shown by the arrows in *Fig 5*. Because the *n*-type material has lost electrons, it acquires a positive potential with respect to the *p*-type material and thus tends to prevent further movement of electrons. The *p*-type material has gained electrons and becomes negatively charged with respect to the *n*-type material and hence tends to retain holes. Thus after a short while, the movement of electrons and holes stops due to the potential difference across the junction, called the **contact potential**. The area in the region of the junction becomes depleted of holes and electrons due to electron-hole recombinations, and is called a **depletion layer**, as shown in *Fig 6*.

7   When an external voltage is applied to a *p–n* junction making the *p*-type material positive with respect to the *n*-type material, as shown in *Fig 7*, the *p–n* junction is **forward biased**. The applied voltage opposes the contact potential, and, in effect, closes the depletion layer. Holes and electrons can now cross the junction and a current flows.

   An increase in the applied voltage above that required to narrow the depletion layer (about 0.2 V for germanium and 0.6 V for silicon), results in a rapid rise in the current flow. Graphs depicting the current-voltage relationship for forward biased *p–n* junctions, for both germanium and silicon, called the forward characteristics, are shown in *Fig 8*.

   When an external voltage is applied to a *p–n* junction making the *p*-type material negative with respect to the *n*-type material, as shown in *Fig 9*, the *p–n* junction is **reverse biased**. The applied voltage is now in the same sense as the contact potential and opposes the movement of holes and electrons, due to opening up the depletion layer. Thus, in theory, no current flows. However, at normal room temperature, certain electrons in the covalent bond lattice acquire sufficient energy from the heat available to leave the lattice, generating mobile electrons and holes. This process is called electron-hole generation by thermal excitation.

   The electrons in the *p*-type material and holes in the *n*-type material caused by thermal excitation, are called **minority carriers** and these will be attracted by the applied voltage. Thus, in practice, a small current of a few μA for germanium and less than one microampere for silicon, at normal room temperature, flows under reverse bias conditions. Typical reverse characteristics are shown in *Fig 10* for both germanium and silicon.

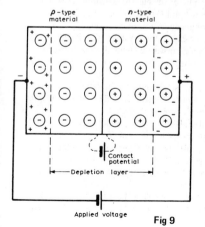

Fig 9

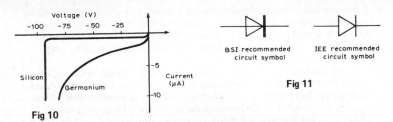

Fig 10

Fig 11

8 **A semiconductor diode** is a device having a *p–n* junction, mounted in a container, suitable for conducting and dissipating the heat generated in operation and having connecting leads. Its operating characteristics are as shown in *Figs 8 and 10*. Two circuit symbols for semiconductor diodes are in common use and are shown in *Fig 11*.

B. WORKED PROBLEMS ON SEMICONDUCTOR DIODES

Problem 1 Explain briefly the terms given below when they are associated with a *p–n* junction:
(a) conduction in intrinsic semiconductors;
(b) majority and minority carriers, and
(c) diffusion.

(a) Silicon or germanium with no doping atoms added are called intrinsic semiconductors. At room temperature, some of the electrons acquire sufficient energy for them to break the covalent bond between atoms and become free mobile electrons. This is called thermal generation of electron-hole pairs. Electrons generated thermally create a gap in the crystal structure called a hole, the atom associated with the hole being positively charged, since it has lost an electron. This positive charge may attract another electron released from another atom, creating a hole elsewhere.

When a potential is applied across the semiconductor material, holes drift towards the negative terminal (unlike charges attract), and electrons towards the positive terminal, and hence a small current flows.

(b) When additional mobile electrons are introduced by doping a semiconductor material with pentavalent atoms (atoms having five valency electrons), these mobile electrons are called majority carriers. The relatively few holes in the *n*-type material produced by intrinsic action are called minority carriers.

For *p*-type materials, the additional holes are introduced by doping with trivalent atoms (atoms having three valency electrons). The holes are apparently positive mobile charges and are majority carriers in the *p*-type material. The relatively few mobile electrons in the *p*-type material produced by intrinsic action are called minority carriers.

(c) Mobile holes and electrons wander freely within the crystal lattice of a semi-conductor material. There are more free electrons in *n*-type material than holes and more holes in *p*-type material than electrons. Thus, in their random wanderings, on average, holes pass into the *n*-type material and electrons into the *p*-type material. This process is called diffusion.

Problem 2 Explain briefly why a junction between *p*-type and *n*-type materials creates a contact potential.

Intrinsic semiconductors have resistive properties, in that when an applied voltage across the material is reversed in polarity, a current of the same magnitude flows in the opposite direction. When a *p–n* junction is formed, the resistive property is replaced by a rectifying property, that is, current passes more easily in one direction than the other.

An *n*-type material can be considered to be a stationary crystal matrix of fixed positive charges together with a number of mobile negative charge carriers (electrons). The total number of positive and negative charges are equal. A *p*-type material can be considered to be a number of stationary negative charges together with mobile positive charge carriers (holes).

Again, the total number of positive and negative charges are equal and the material is neither positively nor negatively charged. When the materials are brought together, some of the mobile electrons in the *n*-type material diffuse into the *p*-type material. Also, some of the mobile holes in the *p*-type material diffuse into the *n*-type material.

Many of the majority carriers in the region of the junction combine with the opposite carriers to complete covalent bonds and create a region on either side of the junction with very few carriers. This region, called the depletion layer, acts as an insulator and is in the order of 0.5 μm thick. Since the *n*-type material has lost electrons, it becomes positively charged. Also, the *p*-type material has lost holes and becomes negatively charged, creating a potential across the junction, called the barrier or contact potential.

Problem 3 Sketch the forward and reverse characteristics of a silicon *p–n* junction diode and describe the shapes of the characteristics drawn.

A typical characteristic for a silicon *p–n* junction having a forward bias is shown in *Fig 8* and having a reverse bias in *Fig 10*. When the positive terminal of the battery is connected to the *p*-type material and the negative terminal to the *n*-type material, the diode is forward biased.

Due to like charges repelling, the holes in the *p*-type material drift towards the junction. Similarly the electrons in the *n*-type material are repelled by the negative bias voltage and also drift towards the junction. The width of the depletion layer and size of the contact potential are reduced. For applied voltages from 0 to about 0.6 V, very little current flows. At about 0.6 V, majority carriers begin to cross the junction in large numbers and current starts to flow. As the applied voltage is raised above 0.6 V, the current increases exponentially (see *Fig 8*).

When the negative terminal of the battery is connected to the *p*-type material and the positive terminal to the *n*-type material, the diode is reverse biased. The holes in the *p*-type material are attracted towards the negative terminal and the electrons in the *n*-type material are attracted towards the positive terminal (unlike charges attract). This drift increases the magnitude of both the contact potential and the thickness of the depletion layer, so that only very few majority carriers have sufficient energy to surmount the junction.

The thermally excited minority carriers, however, can cross the junction since it is, in effect, forward biased for these carriers. The movement of minority carriers results in a small constant current flowing. As the magnitude of the reverse voltage is increased a point will be reached where a large current suddenly starts to flow. The voltage at which this occurs is called the breakdown voltage. This current is due to two effects:
 (i) the **zener effect**, resulting from the applied voltage being sufficient to break some of the covalent bonds, and
 (ii) the **avalanche effect**, resulting from the charge carriers moving at sufficient speed to break covalent bonds by collision.

C. FURTHER PROBLEMS ON SEMICONDUCTOR DIODES

(a) SHORT ANSWER PROBLEMS

1 A good conductor has a resistivity in the order of .... to .... ohm metres.
2 A semiconductor has a resistivity in the order of .... to .... ohm metres.
3 An insulator has a resistivity in the order of .... to .... ohm metres.
4 Over a limited range, the resistance of an insulator .... with increase in temperature.
5 Over a limited range, the resistance of a semiconductor............ with increase in temperature.
6 Over a limited range, the resistance of a conductor .... with increase in temperature.
7 Name two semiconductor materials used in the electronics industry.
8 Name two insulators used in the electronics industry.
9 Name two good conductors used in the electronics industry.
10 The working temperature of germanium should not exceed ....°C to ....°C, depending on its ....
11 The working temperature of silicon should not exceed ....°C to ....°C, depending on its ....
12 Antimony is called .... impurity.
13 Arsenic has .... valency electrons.
14 When phosphorus is introduced into a semiconductor material, mobile .... result.
15 Boron is called a .... impurity.
16 Indium has .... valency electrons.
17 When aluminium is introduced into a semiconductor material, mobile .... result.

18 When a p–n junction is formed, the n-type material acquires a .... charge due to losing ....
19 When a p–n junction is formed, the p-type material acquires a .... charge due to losing ....
20 To forward bias a p–n junction, the .... terminal of the battery is connected to the p-type material.
21 To reverse bias a p–n junction, the positive terminal of the battery is connected to the .... material.
22 When a germanium p–n junction is forward biased, approximately .... mV must be applied before an appreciable current starts to flow.
23 When a silicon p–n junction is forward biased, approximately .... mV must be applied before an appreciable current starts to flow.
24 When a p–n junction is reversed biased, the thickness or width of the depletion layer ....
25 If the thickness or width of a depletion layer decreases, then the p–n junction is .... biased.

(b) MULTI-CHOICE PROBLEMS (Answers on page 152)

In problems 1 to 5, select which statements are true.
1 In pure silicon:
  (a) the holes are the majority carriers;
  (b) the electrons are the majority carriers;
  (c) the holes and electrons exist in equal numbers;
  (d) conduction is due to there being more electrons than holes.
2 Intrinsic semiconductor materials have:
  (a) covalent bonds forming a tetrahedral structure;
  (b) pentavalent atoms added;
  (c) conduction by means of doping;
  (d) a resistance which increases with increase of temperature.
3 Pentavalent impurities:
  (a) have three valency electrons;
  (b) introduce holes when added to a semiconductor material;
  (c) are introduced by adding aluminium atoms to a semiconductor material;
  (d) increase the conduction of a semiconductor material.
4 Free electrons in a p-type material:
  (a) are majority carriers;
  (b) take no part in conduction;
  (c) are minority carriers;
  (d) exist in the same numbers as holes.
5 When an unbiased p–n junction is formed:
  (a) the p-side is positive with respect to the n-side;
  (b) a contact potential exists;
  (c) electrons diffuse from the p-type material to the n-type material;
  (d) conduction is by means of majority carriers.

In *Problems 6 to 10*, select which statements are false.
6 (a) The resistance of an insulator remains approximately constant with increase of temperature.
  (b) The resistivity of a good conductor is about 10^7 to 10^8 ohm metres.
  (c) The resistivity of a conductor increases with increase of temperature.
  (d) The resistance of a semiconductor decreases with increase of temperature.

7 Trivalent impurities:
   (a) have three valency electrons;
   (b) introduce holes when added to a semiconductor material;
   (c) can be introduced to a semiconductor material by adding antimony atoms to it;
   (d) increase the conductivity of a semiconductor material when added to it.
8 Free electrons in an n-type material:
   (a) are majority carriers;
   (b) diffuse into the p-type material when a p–n junction is formed;
   (c) as a result of the diffusion process leave the n-type material positively charged;
   (d) exist in the same numbers as the holes in the n-type material.
9 When a germanium p–n junction diode is forward biased:
   (a) current starts to flow in an appreciable amount when the applied voltage is about 600 mV;
   (b) the thickness or width of the depletion layer is reduced;
   (c) the curve representing the current flow is exponential;
   (d) the positive terminal of the battery is connected to the p-type material.
10 When a silicon p–n junction diode is reverse biased:
   (a) a constant current flows over a large range of voltages;
   (b) current flow is due to electrons in the n-type material;
   (c) current type is due to minority carriers;
   (d) the magnitude of the reverse current flow is usually less than 1 μA.

(c) CONVENTIONAL PROBLEMS

1 Explain what you understand by the term intrinsic semiconductor and how an intrinsic semiconductor is turned into either a p-type or an n-type material. Explain also what is meant by minority and majority carriers in an n-type material and state whether the numbers of each of these carriers are affected by temperature.

2 A piece of pure silicon is doped with (a) pentavalent impurity and (b) trivalent impurity. Explain the effect these impurities have on the form of conduction in silicon.

3 With the aid of simple sketches, explain how pure germanium can be treated in such a way that conduction is predominantly due to (a) electrons and (b) holes.

4 Explain the terms given below when used in semiconductor terminology:
   (a) covalent bond; (b) trivalent impurity; (c) pentavalent impurity; (d) electron-hole pair generation.

5 Explain briefly why although both p-type and n-type materials have resistive properties when separate, they have rectifying properties when a junction between them exists.

6 The application of an external voltage to a junction diode can influence the drift of holes and electrons. With the aid of diagrams explain this statement and also how the direction and magnitude of the applied voltage affects the depletion layer.

7 State briefly what you understand by the terms:
   (a) reverse bias; (b) forward bias; (c) contact potential; (d) diffusion, and (e) minority carrier conduction.

8 Explain briefly the action of a p–n junction diode:
   (a) on open-circuit; (b) when provided with a forward bias, and (c) when provided with a reverse bias. Sketch the characteristic curves for both forward and reverse bias conditions.

9 Draw a diagram illustrating the charge situation for an unbiased p–n junction. Explain the change in the charge situation when compared with that in isolated

p-type and n-type materials. Mark on the diagram the depletion layer and the majority carriers in each region.

10 Give an explanation of the principle of operation of a p–n junction as a rectifier. Sketch the current–voltage characteristics showing the approximate values of current and voltage for a silicon junction diode.

9 Transistors

A MAIN POINTS ASSOCIATED WITH TRANSISTORS

1 The **bipolar junction transistor** consists of three regions of semiconductor material. One type is called a p–n–p transistor, in which two regions of p-type material sandwich a very thin layer of n-type material. A second type is called an n–p–n transistor, in which two regions of n-type material sandwich a very thin layer of p-type material. Both of these types of transistors consist of two p–n junctions placed very close to one another in a back-to-back arrangement on a single piece of semiconductor material. Diagrams depicting these two types of transistors are shown in *Fig 1*.

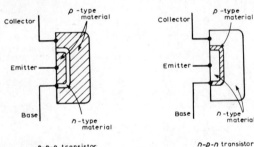

Fig 1

The two p-type material regions of the p–n–p transistor are called the emitter and collector and the n-type material is called the base. Similarly, the two n-type material regions of the n–p–n transistor are called the emitter and collector and the p-type material region is called the base, as shown in *Fig 1*.

2 Transistors have three connecting leads and in operation an electrical input to one pair of connections, say the emitter and base connections can control the output from another pair, say the collector and emitter connections. This type of operation is achieved by appropriately biasing the two internal p–n junctions. When batteries and resistors are connected to a p–n–p transistor, as shown in *Fig 2(a)*, the base-emitter junction is **forward biased** and the base-collector junction is **reverse biased**. Similarly, an n–p–n transistor has its base-emitter junction forward biased and its base-collector junction reverse biased when the batteries are connected as shown in *Fig 2(b)*.

3 For a silicon p–n–p transistor, biased as shown in *Fig 2(a)*, if the base-emitter junction is considered on its own, it is forward biased and a current flows. This is depicted in *Fig 3(a)*. For example, if R_E is 1000 Ω, the battery is 4.5 V and the

128

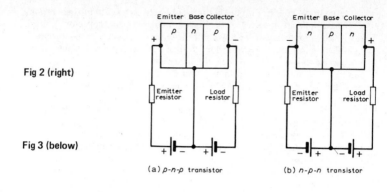

Fig 2 (right)

Fig 3 (below)

(a) *p-n-p* transistor

(b) *n-p-n* transistor

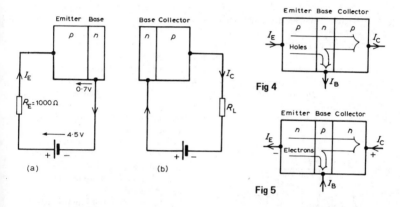

Fig 4

Fig 5

voltage drop across the junction is taken as 0.7 V, the current flowing is given by $\frac{4.5-0.7}{1000} = 3.8$ mA.

When the base–collector junction is considered on its own, as shown in *Fig 3(b)*, it is reverse biased and the collector current is something less than 1 μA.

However, when both external circuits are connected to the transistor, most of the 3.8 mA of current flowing in the emitter, which previously flowed from the base connection, now flows out through the collector connection due to transistor action.

4 In a *p–n–p* transistor, connected as shown in *Fig 2(a)*, transistor action is accounted for as follows:

(a) The majority carriers in the emitter *p*-type material are holes.
(b) The base–emitter junction is forward biased to the majority carriers and the holes cross the junction and appear in the base region.
(c) The base region is very thin and is only lightly doped with electrons so although some electron-hole pairs are formed, many holes are left in the base region.
(d) The base–collector junction is reverse biased to electrons in the base region and

holes in the collector region, but forward biased to holes in the base region. These holes are attracted by the negative potential at the collector terminal.

(e) A large proportion of the holes in the base region cross the base–collector junction into the collector region, creating a collector current. Conventional current flow is in the direction of hole movement.

The transistor action is shown diagrammatically in *Fig 4*. For transistors having very thin base regions, up to 99.5% of the holes leaving the emitter cross the base collector junction.

5   In an *n–p–n* transistor, connected as shown in *Fig 2(b)*, transistor action is accounted for as follows:

(a) The majority carriers in the *n*-type emitter material are electrons.
(b) The base–emitter junction is forward biased to these majority carriers and electrons cross the junction and appear in the base region.
(c) The base region is very thin and only lightly doped with holes, so some recombination with holes occurs but many electrons are left in the base region.
(d) The base–collector junction is reverse biased to holes in the base region and electrons in the collector region, but is forward biased to electrons in the base region. These electrons are attracted by the positive potential at the collector terminal.
(e) A large proportion of the electrons in the base region cross the base-collector junction into the collector region, creating a collector current.

The transistor action is shown diagrammatically in *Fig 5*. As stated in paragraph 4, conventional current flow is taken to be in the direction of hole flow, that is, in the opposite direction to electron flow, hence the directions of the conventional current flow are as shown in *Fig 5*.

6   For a *p–n–p* transistor, the base-collector junction is reverse biased for majority carriers. However, a small leakage current, I_{CBO} flows from the base to the collector due to thermally generated minority carriers (electrons in the collector and holes in the base), being present.

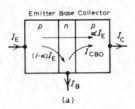

The base-collector junction is forward biased to these minority carriers. If a proportion, α, (having a value of up to 0.995 in modern transistors), of the holes passing into the base from the emitter, pass through the base-collector junction, then the various currents flowing in a *p–n–p* transistor are as shown in *Fig 6(a)*.

Similarly, for an *n–p–n* transistor, the base-collector junction is reverse biased for majority carriers, but a small leakage current, I_{CBO}, flows from the collector to the base due to thermally generated minority carriers (holes in the collector and electrons in the base), being present. The base-collector junction is forward biased to these minority carriers. If a proportion,

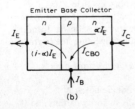

Fig 6

α, of the electrons passing through the base-emitter junction also pass through the base-collector junction, then the currents flowing in an n–p–n transistor are as shown in *Fig 6(b)*.

7 Symbols are used to represent p–n–p and n–p–n transistors in circuit diagrams and two of these in common use are shown in *Fig 7*. The arrow head drawn on the emitter of the symbol is in the direction of conventional emitter current (hole flow). The potentials marked at the collector, base and emitter are typical values for a silicon transistor having a potential difference of 6 V between its collector and its emitter.

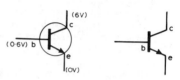

Fig 7

The voltage of 0.6 V across the base and emitter is that required to reduce the potential barrier and if it is raised slightly to, say, 0.62 V, it is likely that the collector current will double to about 2 mA. Thus a small change of voltage between the emitter and the base can give a relatively large change of current in the emitter circuit. Because of this, transistors can be used as amplifiers.

8 There are three ways of connecting a transistor, depending on the use to which it is being put. The ways are classified by the electrode which is common to both the input and the output. They are called:

(a) common-base configuration, shown in *Fig 8(a)*;
(b) common-emitter configuration, shown in *Fig 8(b)*;
(c) common-collector configuration, shown in *Fig 8(c)*.

These configurations are for an n–p–n transistor. The current flows shown are all reversed for a p–n–p transistor.

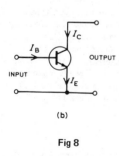

Fig 8

9 The effect of changing one or more of the various voltages and currents associated with a transistor circuit can be shown graphically and these graphs are called the characteristics of the transistor. As there are five variables (collector, base and

emitter currents and voltages across the collector and base and emitter and base) and also three configurations, many characteristics are possible. Some of the possible characteristics are given below.

(a) COMMON-BASE CONFIGURATION

(i) *Input characteristic.* With reference to *Fig 8(a)*, the input to a common-base transistor is the emitter current, I_E, and can be varied by altering the base-emitter voltage V_{EB}. The base-emitter junction is essentially a forward biased junction diode, so as V_{EB} is varied, the current flowing is similar to that for a junction diode, as shown in *Fig 9* for a silicon transistor. *Fig 9* is called the input characteristic for an n–p–n transistor having common-base configuration. The variation of the collector-base voltage V_{CB} has little effect on the characteristic. A similar characteristic can be obtained for a p–n–p transistor, these having reversed polarities.

(ii) *Output characteristics.* The value of the collector current I_C is very largely determined by the emitter current, I_E. For a given value of I_E, the collector-base voltage, V_{CB}, can be varied and has little effect on the value of I_C. If V_{CB} is made slightly negative, the collector no longer attracts the majority carriers leaving the emitter and I_C falls rapidly to zero. A family of curves for various values of I_E are possible and some of these are shown in *Fig 10*. *Fig 10* is called the output characteristics for an n–p–n transistor having common-base configuration. Similar characteristics can be obtained for a p–n–p transistor, these having reversed polarities.

(b) COMMON-EMITTER CONFIGURATION

(i) *Input characteristic.* In a common-emitter configuration (see *Fig 8(b)*), the base current is now the input current. As V_{EB} is varied, the characteristic obtained is similar in shape to the input characteristic for a common-base configuration shown in *Fig 9*, but the values of current are far less. With reference to *Fig 6(a)*, as long as the junctions are biased as described, the three currents I_E, I_C and I_B keep the ratio $1:\alpha:(1-\alpha)$, whichever configuration is adopted. Thus the base current changes are much smaller than the corresponding emitter current changes and the input characteristic for an n–p–n transistor is as shown in *Fig 11*. A similar characteristic can be obtained for a p–n–p transistor, these having reversed polarities.

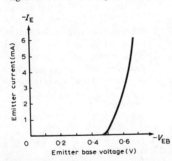

Fig 9

(ii) *Output characteristics.* A family of curves can be obtained, depending on the value of base current I_B and some of these for an n–p–n transistor are shown in *Fig 12*. A similar set of characteristics can be obtained for a p–n–p transistor, these having reversed polarities. These characteristics differ from the common base output characteristics in two ways:

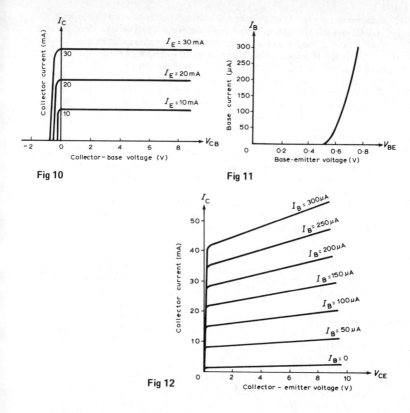

Fig 10

Fig 11

Fig 12

the collector current reduces to zero without having to reverse the collector voltage, and

the characteristics slope upwards indicating a lower output resistance (usually kilohms for a common-emitter configuration compared with megohms for a common-base configuration).

B. WORKED PROBLEMS ON TRANSISTORS

Problem 1 With reference to a *p–n–p* transistor, explain briefly what is meant by the term transistor action and why a bipolar junction transistor is so named.

For the transistor as depicted in *Fig 4*, the emitter is relatively heavily doped with acceptor atoms (holes). When the emitter terminal is made sufficiently positive with respect to the base, the base–emitter junction is forward biased to the majority carriers. The majority carriers are holes in the emitter and these drift from the

133

emitter to the base. The base region is relatively lightly doped with donor atoms (electrons) and although some electron–hole recombinations take place, perhaps 0.5%, most of the holes entering the base, do not combine with electrons.

The base–collector junction is reverse biased to electrons in the base region, but forward biased to holes in the base region. Since the base is very thin and now is packed with holes, these holes pass the base–emitter junction towards the negative potential of the collector terminal. The control of current from emitter to collector is largely independent of the collector–base voltage and almost wholly governed by the emitter–base voltage. The essence of transistor action is this current control by means of the base–emitter voltage.

In a p–n–p transistor, holes in the emitter and collector regions are majority carriers, but are minority carriers when in the base region. Also thermally generated electrons in the emitter and collector regions are minority carriers as are holes in the base region. However, both majority and minority carriers contribute towards the total current flow (see *Fig 6(a)*). It is because a transistor makes use of both types of charge carriers (holes and electrons) that they are called bipolar. The transistor also comprises two p–n junctions and for this reason it is a junction transistor. Hence the name bipolar junction transistor.

Problem 2 The basic construction of an n–p–n transistor makes it appear that the emitter and collector can be interchanged. Explain why this is not usually done.

In principle, a bipolar junction transistor will work equally well with either the emitter or collector acting as the emitter. However, the conventional emitter current largely flows from the collector through the base to the emitter, hence the emitter region is far more heavily doped with donor atoms (electrons) than the base is with acceptor atoms (holes). Also, the base–collector junction is normally reverse biased and in general, doping density increases the electric field in the junction and so lowers the breakdown voltage. Thus, to achieve a high breakdown voltage, the collector region is relatively lightly doped.

In addition, in most transistors, the method of production is to diffuse acceptor and donor atoms onto the n-type semiconductor material, one after the other, so that one overrides the other. When this is done, the doping density in the base region is not uniform but decreases from emitter to collector. This results in increasing the effectiveness of the transistor. Thus, because of the doping densities in the three regions and the non-uniform density in the base, the collector and emitter terminals of a transistor should not be interchanged when making transistor connections.

Problem 3 With the aid of a circuit diagram, explain how the input and output characteristics of an n–p–n transistor having a common–base configuration can be obtained.

A circuit diagram for obtaining the input and output characteristics for an n–p–n transistor connected in common-base configuration is shown in *Fig 13*. The input characteristic can be obtained by varying R_1, which varies V_{EB}, and noting the

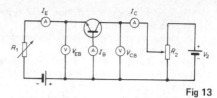

Fig 13

corresponding values of I_E. This is repeated for various values of V_{CB}. It will be found that the input characteristic is almost independent of V_{CB} and it is usual to give only one characteristic, as shown in *Fig 9*.

To obtain the output characteristics, as shown in *Fig 10*, I_E is set to a suitable value by adjusting R_1. For various values of V_{CB}, set by adjusting R_2, I_C is noted. This procedure is repeated for various values of I_E. To obtain the full characteristics, the polarity of battery V_2 has to be reversed to reduce I_C to zero. This must be done very carefully or else values of I_C will rapidly increase in the reverse direction and burn out the transistor.

C. FURTHER PROBLEMS ON TRANSISTORS

(a) SHORT ANSWER PROBLEMS

1 In a p–n–p transistor the p-type material regions are called the .... and ...., and the n-type material region is called the ....
2 In an n–p–n transistor, the p-type material region is called the .... and the n-type material regions are called the .... and the ....
3 In a p–n–p transistor, the base–emitter junction is .... biased and the base–collector junction is .... biased.
4 In an n–p–n transistor, the base–collector junction is .... biased and the base–emitter junction is .... biased.
5 Majority charge carriers in the emitter of a transistor pass into the base region. Most of them do not recombine because the base is .... doped.
6 Majority carriers in the emitter region of a transistor pass the base–collector junction because for these carriers it is .... biased.
7 Conventional current flow is in the direction of .... flow.
8 Leakage current flows from .... to .... in an n–p–n transistor.
9 The input characteristic of I_E against V_{EB} for a transistor in common-base configuration is similar in shape to that of a ....
10 The output resistance of a transistor connected in common-emitter configuration is .... than that of a transistor connected in common-base configuration.

(b) MULTI-CHOICE PROBLEMS (Answers on page 153)

In *Problems 1 to 10* select the correct answer from those given.
1 In normal operation, the junctions of a p–n–p transistor are:
  (a) both forward biased;
  (b) base–emitter forward biased and base–collector reverse biased;
  (c) both reverse biased;
  (d) base–collector forward biased and base–emitter reverse biased.

2. In normal operation, the junctions of an n–p–n transistor are:
   (a) both forward biased;
   (b) base–emitter forward biased and base–collector reverse biased;
   (c) both reverse biased;
   (d) base–collector forward biased and base–emitter reverse biased.

3. The current flow across the base–emitter junction of a p–n–p transistor consists of:
   (a) mainly electrons;
   (b) equal numbers of holes and electrons;
   (c) mainly holes;
   (d) the leakage current.

4. The current flow across the base–emitter junction of an n–p–n transistor consists of:
   (a) mainly electrons;
   (b) equal numbers of holes and electrons;
   (c) mainly holes;
   (d) the leakage current.

5. In normal operation an n–p–n transistor connected in common–base configuration has
   (a) the emitter at a lower potential than the base;
   (b) the collector at a lower potential than the base;
   (c) the base at a lower potential than the emitter;
   (d) the collector at a lower potential than the emitter.

6. In normal operation, a p–n–p transistor connected in common–base configuration has
   (a) the emitter at a lower potential than the base;
   (b) the collector at a higher potential than the base;
   (c) the base at a higher potential than the emitter;
   (d) the collector at a lower potential than the emitter.

7. If the per unit value of electrons which leave the emitter and pass to the collector, α, is 0.9 in an n–p–n transistor and the emitter current is 4 mA, then
   (a) the base current is approximately 4.4 mA;
   (b) the collector current is approximately 3.6 mA;
   (c) the collector current is approximately 4.4 mA;
   (d) the base current is approximately 3.6 mA.

8. The base region of a p–n–p transistor is
   (a) very thin and heavily doped with holes;
   (b) very thin and heavily doped with electrons;
   (c) very thin and lightly doped with holes;
   (d) very thin and lightly doped with electrons.

9. The voltage drop across the base–emitter junction of a p–n–p silicon transistor in normal operation is about
   (a) 200 mV;
   (b) 600 mV;
   (c) zero;
   (d) 4.4 V.

10. For a p–n–p transistor,
    (a) the number of majority carriers crossing the base–emitter junction largely depends on the collector voltage;
    (b) in common–base configuration, the collector current is proportional to the collector–base voltage;
    (c) in common–emitter configuration, the base current is less than the base current in common–base configuration;

(d) the collector current flow is independent of the emitter current flow for a given value of collector–base voltage.

(c) CONVENTIONAL PROBLEMS

1. Explain with the aid of sketches, the operation of an n–p–n transistor and also explain why the collector current is very nearly equal to the emitter current.
2. Explain what is meant by the term 'transistor action'.
3. Describe the basic principle of operation of a bipolar junction transistor including why majority carriers crossing into the base from the emitter pass to the collector and why the collector current is almost unaffected by the collector potential.
4. For a transistor connected in common–emitter configuration, sketch the output characteristics relating collector current and the collector–emitter voltage, for various values of base current. Explain the shape of the characteristics.
5. Sketch the input characteristic relating emitter current and the emitter–base voltage for a transistor connected in common–base configuration, and explain its shape.
6. With the aid of a circuit diagram, explain how the output characteristics of an n–p–n transistor having common–base configuration may be obtained and any special precautions which should be taken.
7. Draw sketches to show the direction of the flow of leakage current in both n–p–n and p–n–p transistors. Explain the effect of leakage current on a transistor connected in common–base configuration.
8. Using the circuit symbols for transistors show how (a) common–base, and (b) common–emitter configuration can be achieved. Mark on the symbols the inputs, the outputs, polarities under normal operating conditions to give correct biasing and current directions.
9. Draw a diagram showing how a transistor can be used in common emitter configuration. Mark on the sketch the input, output, polarities under normal operating conditions and current directions.
10. Sketch the circuit symbols for (a) a p–n–p and (b) an n–p–n transistor. Mark on the emitter electrodes the direction of conventional current flow and explain why the current flows in the direction indicated.

10 Measuring instruments and measurements

A. MAIN POINTS RELATING TO MEASURING INSTRUMENTS AND MEASUREMENTS

1 Tests and measurements are important in designing, evaluating, maintaining and servicing electrical circuits and equipment. In order to detect electrical quantities such as current, voltage, resistance or power, it is necessary to transform an electrical quantity or condition into a visible indication. This is done with the aid of instruments (or meters) that indicate the magnitude of quantities either by the position of a pointer moving over a graduated scale (called an **analogue** instrument or in the form of a decimal number (called a **digital** instrument).

2 All analogue electrical indicating instruments require three essential devices. These are:
  (a) **A deflecting or operating device.** A mechanical force is produced by the current or voltage which causes the pointer to deflect from its zero position.
  (b) **A controlling device.** The controlling force acts in opposition to the deflecting force and ensures that the deflection shown on the meter is always the same for a given measured quantity. It also prevents the pointer always going to the maximum deflection. There are two main types of controlling device—spring control and gravity control.
  (c) **A damping device.** The damping force ensures that the pointer comes to rest in its final position quickly and without undue oscillation. There are three main types of damping used—eddy-current damping, air-friction damping and fluid-friction damping.

3 There are basically two types of scale—linear and non-linear.
A **linear scale** is shown in *Fig 1(a)* where each of the divisions or graduations are evenly spaced. The voltmeter shown has a range 0–100 V, i.e. a full scale deflection (FSD) of 100 V.

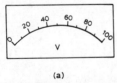
(a)

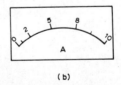
(b)

Fig 1

A **non-linear scale** is shown in *Fig 1(b)*. The scale is cramped at the beginning and the graduations are uneven throughout the range. The ammeter shown has a FSD of 10 A.

4 Comparison of the moving coil, moving iron and moving coil rectifier instruments (see table opposite).

(For the principle of operation of moving coil and moving iron instruments, see worked problems 1 and 2).

138

Type of instrument	Moving coil	Moving iron	Moving coil rectifier
Suitable for measuring	Direct current and voltage	Direct and alternating current and voltage (reading in rms value)	Alternating current and voltage (reads average value but scale is adjusted to give rms value for sinusoidal waveforms)
Scale	Linear	Non-linear	Linear
Method of control	Hairsprings	Hairsprings	Hairsprings
Method of damping	Eddy current	Air	Eddy current
Frequency limits	–	20–200 Hz	20–100 kHz
Advantages	1 Linear scale 2 High sensitivity 3 Well shielded from stray magnetic fields 4 Lower power consumption	1 Robust construction 2 Relatively cheap 3 Measures dc and ac 4 In frequency range 20–100 Hz reads rms correctly, regardless of supply waveform	1 Linear scale 2 High sensitivity 3 Well shielded from stray magnetic fields 4 Low power consumption 5 Good frequency range
Disadvantages	1 Only suitable for dc 2 More expensive than moving iron type 3 Easily damaged	1 Non-linear scale 2 Affected by stray magnetic fields 3 Hysteresis errors in dc circuits 4 Liable to temperature errors 5 Due to the inductance of the solenoid, readings can be affected by variation of frequency	1 More expensive than moving iron type 2 Errors caused when supply is non-sinusoidal

5. A moving coil instrument, which measures only dc, may be used in conjunction with a bridge rectifier circuit as shown in *Fig 2* to provide an indication of alternating currents and voltages. The average value of the full wave rectified current is $0.637\,I_M$. However, a meter being used to measure ac is usually calibrated in rms values. For sinusoidal quantities the indication is

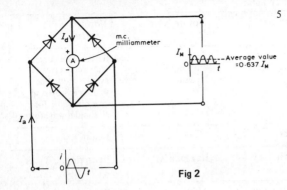

Fig 2

$\dfrac{0.707\,I_M}{0.637\,I_M}$, i.e. 1.11 times the mean value.

Rectifier instruments have scaled calibrated in rms quantities and it is assumed by the manufacturer that the ac is sinusoidal.

6. An **ammeter**, which measures current, has a low resistance (ideally zero) and must be connected in series with the circuit.

7. A **voltmeter**, which measures pd, has a high resistance (ideally infinite) and must be connected in parallel with the part of the circuit whose pd is required.

8. There is no difference between the basic instrument used to measure current and voltage since both use a milliammeter as their basic part. This is a sensitive instrument which gives FSD for currents of only a few milliamperes. When an ammeter is required to measure currents of larger magnitude, a proportion of the current is diverted through a low value resistance connected in parallel with the meter. Such a diverting resistor is called a **shunt**.

From *Fig 3(a)*, $V_{PQ} = V_{RS}$. Hence $I_a r_a = I_S R_S$

Thus the value of the shunt, $R_S = \dfrac{I_a r_a}{I_s}\ \Omega$

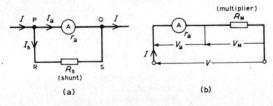

Fig 3

The milliammeter is converted into a voltmeter by connecting a high resistance (called a **multiplier**) in series with it as shown in *Fig 3(b)*.

From *Fig 3(b)*, $V = V_a + V_M = I r_a + I R_M$

Thus the value of the multiplier, $R_M = \dfrac{V - I r_a}{I}\ \Omega$

9. An **ohmmeter** is an instrument for measuring electrical resistance. A simple

ohmmeter circuit is shown in *Fig 4(a)*. Unlike the ammeter or voltmeter, the ohmmeter circuit does not receive the energy necessary for its operation from the circuit under test. In the ohmmeter this energy is supplied by a self-contained source of voltage, such as a battery. Initially, terminals XX are short-circuited and R adjusted to give FSD on the milliammeter. If current I is at a maximum value and voltage E is constant, then resistance $R = E/I$ is at a minimum value. Thus FSD on the milliammeter is made zero on the resistance scale. When terminals XX are open circuited no

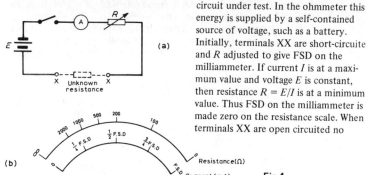

Fig 4

current flows and $R(= E/0)$ is infinity, ∞. The milliammeter can thus be calibrated directly in ohms. A cramped (non-linear) scale results and is 'back to front' (as shown in *Fig 4(b)*). When calibrated, an unknown resistance is placed between terminals XX and its value determined from the position of the pointer on the scale. An ohmmeter designed for measuring low values of resistance is called a **continuity tester**. An ohmmeter designed for measuring high values of resistance (i.e. megohms) is called an **insulation resistance tester** (or **'Megger'**).

10 Instruments are manufactured that combine a moving coil meter with a number of shunts and series multipliers, to provide a range of readings on a single scale graduated to read current and voltage. If a battery is incorporated into the instrument then resistance can also be measured. Such instruments are called **multimeters** or **universal instruments** or **multirange instruments**. An 'Avometer' is a typical example. A particular range may be selected either by the use of separate terminals or by a selector switch. Only one measurement can be performed at one time. Often such instruments can be used in ac as well as dc circuits when a rectifier is incorporated in the instrument.

11 A **wattmeter** is an instrument for measuring electrical power in a circuit. *Fig 5* shows typical connections of a wattmeter used for measuring power supplied to a load.

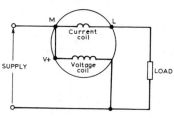

The instrument has two coils:
(i) a current coil, which is connected in series with the load (like an ammeter), and
(ii) a voltage coil, which is connected in parallel with the load (like a voltmeter).

Fig 5

12 The **cathode ray oscilloscope (CRO)** may be used in the observation of waveforms and for the measurement of voltage, current, frequency, phase and time.
   (i) With **direct voltage measurements**, only the Y amplifier 'volts/cm' switch on

the CRO is used. With no voltage applied to the Y plates the position of the spot trace on the screen is noted. When a direct voltage is applied to the Y plates the new position of the spot trace is an indication of the magnitude of the voltage. For example, in *Fig 6(a)*, with no voltage applied to the Y plates, the spot trace is in the centre of the screen (initial position) and then the spot trace moves 2.5 cm to the final position shown, on application of a dc voltage. With the 'volts/cm' switch on 10 volts/cm the magnitude of the direct voltage is 2.5 cm × 10 V/cm, i.e. 25 V.

(ii) With **alternating voltage measurement**, let a sinusoidal waveform be displayed on a CRO screen as shown in *Fig 6(b)*. If the 'variable' switch is on, say 5 ms/cm then the **periodic time** T of the sine wave is 5 ms/cm × 4 cm i.e. 20 ms or 0.02 s.

Since frequency $f = \dfrac{1}{T}$,

frequency $= \dfrac{1}{0.02} = 50$ **Hz**

If the 'volts/cm' switch is on, say, 20 volts/cm then the **amplitude or peak value** of the sine wave shown is 20 volts/cm × 2 cm i.e. **40 Volts**.

Since rms voltage $= \dfrac{\text{peak voltage}}{\sqrt{2}}$

rms voltage $= \dfrac{40}{\sqrt{2}} = 28.28$ **V**.

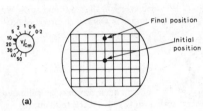

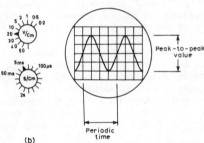

Fig 6

13 An **electronic voltmeter** can be used to measure with accuracy emf or pd from mV to kV by incorporating in its design amplifiers and attenuators.

14 A **null method of measurement** is a simple, accurate and widely used method which depends on an instrument reading being adjusted to read zero current only. The method assumes:

(i) if there is any deflection at all, then some current is flowing, and

(ii) if there is no deflection, then no current flows (i.e. a null condition).

Hence it is unnecessary for a meter sensing current flow to be calibrated when used in this way. A sensitive milliammeter or microammeter with centre zero position setting is called a **galvanometer**. Two examples where the method is used are in the Wheatstone bridge and in the dc potentiometer.

15 *Fig 7* shows a **Wheatstone bridge** circuit which compares an unknown resistance R_x with others of known values, i.e. R_1 and R_2 which have fixed values and R_3 which is variable. R_3 is varied until zero deflection is obtained on the galvano-

meter G. No current then flows through the meter, $V_A = V_B$, and the bridge is said to be 'balanced'.

At balance, $R_1 R_x = R_2 R_3$,

i.e. $R_x = \dfrac{R_2 R_3}{R_1}\ \Omega$

16 The **dc potentiometer** is a null-balance instrument used for determining values of emf's and pd's by comparison with a known emf or pd. In *Fig 8(a)*, using a standard cell of known emf E_1, the slider S is moved along the slide wire until balance is obtained (i.e. the galvanometer deflection is zero), shown as length l_1. The standard cell is now replaced by a cell of unknown emf, E_2 (see *Fig 8(b)*) and again balance is obtained (shown as l_2). Since $E_1 \propto l_1$ and $E_2 \propto l_2$, then

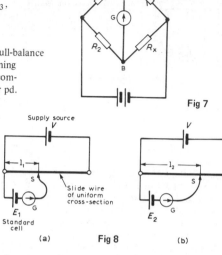

Fig 7

Fig 8

$\dfrac{E_1}{E_2} = \dfrac{l_1}{l_2}$ and $E_2 = E_1 \left(\dfrac{l_2}{l_1}\right)$ **volts.**

17 The errors most likely to occur in measurements are those due to
   (i) the limitations of the instrument;
   (ii) the operator; and
   (iii) the instrument disturbing the circuit.
   (see *Problem 9*).

B. WORKED PROBLEMS ON MEASURING INSTRUMENTS AND MEASUREMENTS

Problem 1 Describe, with the aid of diagrams, the principle of operation of a moving coil instrument.

A moving coil instrument operates on the motor principle. When a conductor carrying current is placed in a magnetic field, a force F is exerted on the conductor, given by $F = BIl$. If the flux density B is made constant (by using permanent magnets) and the conductor is a fixed length (say, a coil) then the force will depend only on the current flowing in the conductor.

In a moving coil instrument, a coil is placed centrally in the gap between shaped pole pieces as shown by the front elevation in *Fig 9(a)*. The coil is supported by steel pivots, resting in jewel bearings, on a cylindrical iron core. Current is led

143

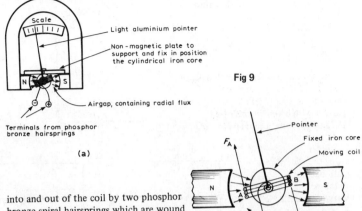

Fig 9

into and out of the coil by two phosphor bronze spiral hairsprings which are wound in opposite directions to minimise the effect of temperature change and to limit the coil swing (i.e. to **control** the movement) and return the movement to zero position when no current flows.

Current flowing in the coil produces forces as shown in *Fig 9(b)*, the directions being obtained by Fleming's left-hand rule. The two forces, F_A and F_B, produce a torque which will move the coil in a clockwise direction, i.e. move the pointer from left to right. Since force is proportional to current the scale is linear.

When the aluminium frame, on which the coil is wound, is rotated between the poles of the magnet, small currents (called eddy currents) are induced into the frame, and this provides automatically the necessary **damping** of the system due to the reluctance of the former to move within the magnetic field.

The moving coil instrument will only measure direct current or voltage and the terminals are marked positive and negative to ensure that the current passes through the coil in the correct direction to deflect the pointer 'up the scale'.

The range of this sensitive instrument is extended by using shunts and multipliers.

Problem 2 Draw diagrams to represent (a) the attraction type, and (b) the repulsion type of the moving iron instrument and briefly describe, for each, their principle of operation.

(a) An **attraction type** of moving iron instrument is shown diagrammatically in *Fig 10(a)*. When current flows in the solenoid, a pivoted soft iron disc is attracted towards the solenoid and the movement causes a pointer to move across a scale.
(b) In the **repulsion type** moving iron instrument shown diagrammatically in *Fig 10(b)*, two pieces of iron are placed inside the solenoid, one being fixed, and

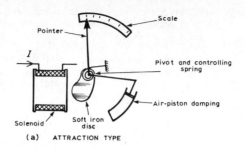

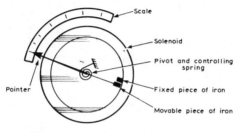

Fig 10   (b)   REPULSION TYPE

the other attached to the spindle carrying the pointer. When current passes through the solenoid, the two pieces of iron are magnetised in the same direction and therefore repel each other. The pointer thus moves across the scale.

The force moving the pointer is, in each type, proportional to I^2. Because of this the direction of current does not matter and the moving iron instrument can be used on dc or ac. The scale, however, is non-linear.

Problem 3 A moving coil instrument gives a FSD when the current is 40 mA and its resistance is 25 Ω. Calculate the value of the shunt to be connected in parallel with the meter to enable it to be used as an ammeter for measuring currents up to 50 A.

The circuit diagram is shown in *Fig 11*, where r_a = resistance of instrument = 25 Ω;

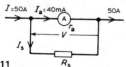

Fig 11

R_s = resistance of shunt;
I_a = maximum permissible current flowing in instrument = 40 mA = 0.04 A;
I_s = current flowing in shunt;
I = total circuit current required to give FSD = 50 A.

Since $I = I_a + I_s$ then $I_s = I - I_a = 50 - 0.04 = 49.96$ A.
$V = I_a r_a = I_s R_s$

Hence $R_s = \dfrac{I_a r_a}{I_s} = \dfrac{(0.04)(25)}{49.96} = 0.020\,02\ \Omega = \mathbf{20.02\ m\Omega}$

Thus for the moving coil instrument to be used as an ammeter with a range 0–50 A, a resistance of value 20.02 mΩ needs to be connected in parallel with the instrument.

Problem 4 A moving coil instrument having a resistance of 10Ω, gives a FSD when the current is 8 mA. Calculate the value of the multiplier to be connected in series with the instrument so that it can be used as a voltmeter for measuring pd's up to 100 V.

The circuit diagram is shown in *Fig 12*, where r_a = resistance of instrument = 10 Ω;

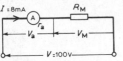

Fig 12

R_M = resistance of multiplier;
I = total permissible instrument current = 8 mA = 0.008 A;
V = total pd required to give FSD = 100 V.

$V = V_a + V_M = Ir_a + IR_M$
i.e. $100 = (0.008)(10) + (0.008)R_M$
$100 - 0.08 = 0.008 R_M$

$R_M = \dfrac{99.92}{0.008} = 12\,490\ \Omega = \mathbf{12.49\ k\Omega}$

Hence for the moving coil instrument to be used as a voltmeter with a range 0–100 V, a resistance of value 12.49 kΩ needs to be connected in series with the instrument.

Problem 5 Calculate the power dissipated by the voltmeter and by resistor R in *Fig 13* when (a) $R = 250\ \Omega$, and (b) $R = 2\ \text{M}\Omega$. Assume that the voltmeter sensitivity (sometimes called the figure of merit) is 10 kΩ/V.

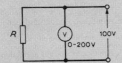

Fig 13

(a) Resistance of voltmeter, R_v = sensitivity × FSD
Hence, $R_v = 10\ \text{k}\Omega/\text{V} \times 200\ \text{V} = 2000\ \text{k}\Omega = 2\ \text{M}\Omega$

Current flowing in voltmeter, $I_v = \dfrac{V}{R_v} = \dfrac{100}{2 \times 10^6} = 50 \times 10^{-6}$ A

Power dissipated by voltmeter = $VI_v = (100)(50 \times 10^{-6}) = \mathbf{5\ mW}$

When $R = 250\ \Omega$, current in resistor, $I_R = \dfrac{V}{R} = \dfrac{100}{250} = 0.4$ A

Power dissipated in load resistor $R = VI_R = (100)(0.4) = \mathbf{40\ W}$

Thus the power dissipated in the voltmeter is insignificant in comparison with the power dissipated in the load.

(b) When $R = 2\text{M}\Omega$, current in resistor, $I_R = \dfrac{100}{2 \times 10^6} = 50 \times 10^{-6}$ A

Power dissipated in load resistor $R = VI_R = 100 \times 50 \times 10^{-6} = \mathbf{5\ mW}$

In this case the higher load resistance reduced the power dissipated such that the voltmeter is using as much power as the load.

Problem 6 An ammeter has a FSD of 100 mA and a resistance of 50 Ω. The ammeter is used to measure the current in a load of resistance 500 Ω when the supply voltage is 10 V. Calculate (a) the ammeter reading expected (neglecting its resistance); (b) the actual current in the circuit; (c) the power dissipated in the ammeter; and (d) the power dissipated in the load.

(a) Expected ammeter reading

$$= \frac{V}{R} = \frac{10}{500} = 20 \text{ mA}$$

(b) Actual ammeter reading

$$= \frac{V}{R+r_a} = \frac{10}{500+50} = 18.18 \text{ mA}$$

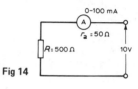

Fig 14

Thus the ammeter itself has caused the circuit conditions to change from 20 mA to 18.18 mA.

(c) Power dissipated in the ammeter = $I^2 R = (18.18 \times 10^{-3})^2 (50) = $ **16.53 mW**

(d) Power dissipated in the load resistor = $I^2 R = (18.18 \times 10^{-3})^2 (500) = $ **165.3 mW**

Problem 7 In a Wheatstone bridge ABCD, a galvanometer is connected between A and C, and a battery between B and D. A resistor of unknown value is connected between A and B. When the bridge is balanced, the resistance between B and C is 100 Ω, that between C and D is 10 Ω and that between D and A is 400 Ω. Calculate the value of the unknown resistance.

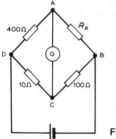

The Wheatstone bridge is shown in *Fig 15* where R_x is the unknown resistance. At balance, equating the products of opposite ratio arms, gives:

$(R_x)(10) = (100)(400)$

$$R_x = \frac{(100)(400)}{10} = 4000 \text{ Ω}$$

Hence unknown resistance, $R_x = $ **4 kΩ**

Fig 15

Problem 8 In a dc potentiometer balance is obtained at a length of 400 mm when using a standard cell of 1.0186 V. Determine the emf of a dry cell if balance is obtained with a length of 650 mm.

$E_1 = 1.0186$ V, $l_1 = 400$ mm, $l_2 = 650$ mm.

With reference to *Fig 8*, $\dfrac{E_1}{E_2} = \dfrac{l_1}{l_2}$

from which $E_2 = E_1 \left(\dfrac{l_2}{l_1}\right) = (1.0186)\left(\dfrac{650}{400}\right) = \mathbf{1.655}$ **V**

Problem 9 List the errors most likely to occur in measurements made with electrical measuring instruments.

(i) Errors in the limitations of the instrument

The calibration accuracy of an instrument depends on the precision with which it is constructed. Every instrument has a margin of error which is expressed as a percentage of the indication. For example, industrial grade instruments have an accuracy of ± 2% of FSD. Thus if a voltmeter has a FSD of 100 V and it indicates 40 V say, then the actual voltage might be anywhere between 40 ± (2% of 100), i.e. 40 ± 2, i.e. between 38 V and 42 V.

When an instrument is calibrated, it is compared against a standard instrument and a graph is drawn of 'error' against 'meter deflection'. A typical graph is shown

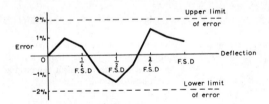

Fig 16

in *Fig 16* where it is seen that the accuracy varies over the scale length. Thus a meter with a ± 2% FSD accuracy would tend to have an accuracy which is much better than ± 2% FSD over much of the range.

(ii) Errors by the operator

It is easy for an operator to misread an instrument reading. With linear scales the values of the sub-divisions are reasonably easy to determine; non-linear scale graduations are more difficult to estimate. Also, scales differ from instrument to instrument and some meters have more than one scale (as with multimeters) and mistakes in reading indications are easily made.

When reading a meter scale it should be viewed from an angle perpendicular to the surface of the scale at the location of the pointer; a meter scale should not be viewed 'at an angle'.

(iii) Errors due to the instrument disturbing the circuit

Any instrument connected into a circuit will affect that circuit to some extent. Meters require some power to operate, but provided this power is small compared with the power in the measured circuit, then little error will result. Incorrect positioning of instruments in a circuit can be a source of errors.

For example, let a resistance be measured by the voltmeter–ammeter method as shown in *Fig 17*. Assuming 'perfect' instruments, the resistance should be given by the voltmeter reading divided by the ammeter reading (i.e. $R = V/I$). However, in *Fig 17(a)*, $V/I = R+r_a$ and in *Fig 17(b)* the current through the ammeter is that through the resistor plus that through the voltmeter. Hence the voltmeter reading divided by the ammeter reading will not give the true value of the resistance R for either methods of connection.

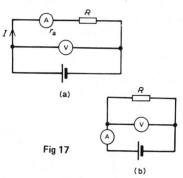

Fig 17

C. FURTHER PROBLEMS ON MEASURING INSTRUMENTS AND MEASUREMENTS

(a) SHORT ANSWER PROBLEMS

1. What is the main difference between an analogue and a digital type measuring instrument?
2. Name the three essential devices for all analogue electrical indicating instruments.
3. Complete the following statements:
   (a) An ammeter has a .... resistance and is connected .... with the circuit.
   (b) A voltmeter has a .... resistance and is connected .... with the circuit.
4. State two advantages and two disadvantages of a moving-coil instrument.
5. What effect does the connection of (a) a shunt; (b) a multiplier; have on a milliammeter?
6. State two advantages and two disadvantages of a moving-iron instrument.
7. Briefly explain the principle of operation of an ohmmeter.
8. Name a type of ohmmeter used for measuring (a) low resistance values; (b) high resistance values.
9. What is a multimeter?
10. When might a rectifier instrument be used in preference to either the moving coil or moving iron instrument?

11 What is the principle of the Wheatstone bridge?
12 How may a dc potentiometer be used to measure pd's?
13 What is meant by a null method of measurement?
14 Define 'calibration accuracy' as applied to a measuring instrument.
15 State three main areas where errors are most likely to occur in measurements.

(b) MULTI-CHOICE PROBLEMS (Answers on page 153)

1 Which of the following would apply to a moving coil instrument?
   (a) An uneven scale, measuring dc; (b) An even scale, measuring ac; (c) An uneven scale, measuring ac; (d) An even scale, measuring dc.
2 In *Problem 1*, which would refer to a moving iron instrument?
3 In *Problem 1*, which would refer to a moving coil rectifier instrument?
4 Which of the following is needed to extend the range of a milliammeter to read voltages of the order of 100 V?
   (a) A parallel high value resistance; (b) A series high value resistance; (c) A parallel low value resistance; (d) A series low value resistance.
5 *Fig 18* shows a scale of a multi-range ammeter. What is the current indicated when switched to a 25 A scale? (a) 84 A; (b) 5.6 A; (c) 14 A; (d) 8.4 A.

Fig 18

A sinusoidal waveform is displayed on a CRO screen. The peak-to-peak distance is 5 cm and the distance between cycles is 4 cm. The 'variable' switch is on 100 μs/cm and the 'volts/cm' switch is on 10 V/cm. In *Problems 6 to 10*, select the correct answer from the following list:

(a) 25 V; (b) 5 V; (c) 0.4 ms; (d) 35.4 V; (e) 4 ms; (f) 50 V; (g) 250 Hz; (h) 2.5 V; (i) 2.5 kHz; (j) 17.7 V.

6 Determine the peak-to-peak voltage.
7 Determine the periodic time of the waveform.
8 Determine the maximum value of voltage.
9 Determine the frequency of the waveform.
10 Determine the rms value of the waveform.

(c) CONVENTIONAL PROBLEMS

1 A moving coil instrument gives FSD for a current of 10 mA. Neglecting the resistance of the instrument, calculate the approximate value of series resistance needed to enable the instrument to measure up to (a) 20 V; (b) 100 V; (c) 250 V.

[(a) 2 kΩ; (b) 10 kΩ; (c) 25 kΩ]

2 A meter of resistance 50 Ω has a FSD of 4 mA. Determine the value of shunt resistance required in order that the FSD should be (a) 15 mA; (b) 20 A; (c) 100 A.

[(a) 18.18 Ω; (b) 10.00 mΩ; (c) 2.00 mΩ]

3 Describe, with the aid of diagrams, the principle of operation of a moving-coil instrument. A moving-coil instrument having a resistance of 20 Ω, gives a FSD when the current is 5 mA. Calculate the value of the multiplier to be connected in series with the instrument so that it can be used as a voltmeter for measuring pd's up to 200 V.

[39.98 kΩ]

4. A moving-coil instrument has a FSD current of 20 mA and a resistance of 25 Ω. Calculate the values of resistance required to enable the instrument to be used (a) as a 0–10 A ammeter, and (b) as a 0–100 V voltmeter. State the mode of resistance connection in each case.

[(a) 50.10 mΩ in parallel; (b) 4.975 kΩ in series]

5. A meter has a resistance of 40 Ω and registers a maximum deflection when a current of 15 mA flows. Calculate the value of resistance that converts the movement into (a) an ammeter with a maximum deflection of 50 A, and (b) a voltmeter with a range of 0–250 V.

[(a) 12.00 mΩ in parallel; (b) 16.63 kΩ in series]

6. (a) Describe, with the aid of diagrams, the principle of operation of a moving-iron instrument.
   (b) Draw a circuit diagram showing how a moving-coil instrument may be used to measure alternating current.
   (c) Discuss the advantages and disadvantages of moving-coil rectifier instruments when compared with moving-iron instruments.

7. (a) Describe, with the aid of a diagram, the principle of the Wheatstone bridge and hence deduce the balance condition giving the unknown resistance in terms of known values of resistance.
   (b) In a Wheatstone bridge PQRS, a galvanometer is connected between Q and S and a voltage source between P and R. An unknown resistor R_x is connected between P and Q. When the bridge is balanced, the resistance between Q and R is 200 Ω, that between R and S is 10 Ω and that between S and P is 150 Ω. Calculate the value of R_x.

[3 kΩ]

8. (a) Describe, with the aid of a diagram, how a dc potentiometer can be used to measure the emf of a cell.
   (b) Balance is obtained in a dc potentiometer at a length of 31.2 cm when using a standard cell of 1.0186 V. Calculate the emf of a dry cell if balance is obtained with a length of 46.7 cm.

[1.525 V]

9. List the errors most likely to occur in the measurements of electrical quantities. A 240 V supply is connected across a load resistance R. Also connected across R is a voltmeter having a FSD of 300 V and a figure of merit (i.e. sensitivity) of 8 kΩ/V. Calculate the power dissipated by the voltmeter and by the load resistance if (a) $R = 100$ Ω; (b) $R = 1$ MΩ. Comment on the results obtained.

[(a) 24 mW; 576 W; (b) 24 mW; 57.6 mW]

10. A 0–1 A ammeter having a resistance of 50 Ω is used to measure the current flowing in a 1 kΩ resistor when the supply voltage is 250 V. Calculate (a) the approximate value of current (neglecting the ammeter resistance), (b) the actual current in the circuit, (c) the power dissipated in the ammeter, and (d) the power dissipated in the 1 kΩ resistor.

[(a) 0.250 A; (b) 0.238 A; (c) 2.832 W; (d) 56.64 W]

Answers to multi-choice problems

CHAPTER 1 (*page 6*)

1 (b); 2 (c); 3 (d); 4 (d); 5 (b); 6 (a); 7 (b); 8 (c).

CHAPTER 2 (*page 25*)

1 (a); 2 (c); 3 (b); 4 (d); 5 (b); 6 (a); 7 (c); 8 (a); 9 (d); 10 (c).

CHAPTER 3 (*page 43*)

1 (b); 2 (a); 3 (b); 4 (c); 5 (a); 6 (b); 7 (b); 8 (c); 9 (c); 10 (c).

CHAPTER 4 (*page 57*)

1 (d); 2 (b); 3 (d); 4 (i); 5 (f); 6 (j); 7 (h); 8 (c); 9 (c); 10 (a) and (d); (b) and (f); (c) and (e).

CHAPTER 5 (*page 73*)

1 (c); 2 (c); 3 (b); 4 (d); 5 (c); 6 (d); 7 (a); 8 (c); 9 (a); 10 (c); 11 (b); 12 (a).

CHAPTER 6 (*page 96*)

1 (c); 2 (d); 3 (d); 4 (a); 5 (d); 6 (c); 7 (b); 8 (c); 9 (b); 10 (c).

CHAPTER 7 (*page 113*)

1 (c); 2 (a); 3 (b); 4 (b); 5 (a); 6 (d); 7 (d); 8 (d); 9 (b); 10 (c); 11 (b); 12 (c).

CHAPTER 8 (*page 125*)

1 (c); 2 (a); 3 (d); 4 (c); 5 (b); 6 (b); 7 (c); 8 (d); 9 (a); 10 (b).

CHAPTER 9 (*page 135*)

1 (b); 2 (b); 3 (c); 4 (a); 5 (a); 6 (d); 7 (b);
8 (d); 9 (b); 10 (c).

CHAPTER 10 (*page 150*)

1 (d); 2 (a) or (c); 3 (b); 4 (b); 5 (c); 6 (f); 7 (c);
8 (a); 9 (i); 10 (j).

Index

Absolute permeability, 49
  permittivity, 32, 33
AC circuit theory, 100, 105
Alternating current, 79
  values, 81, 85
Alternator, 80
Ammeter, 140
Amplitude, 81
Analogue instrument, 138
Apparent power, 104
Avalanche effect, 124
Average value, 81
Avometer, 141

B–H curves, 49, 50

Capacitance, 32, 35
Capacitive reactance, 101, 105
Capacitors, 32
  energy stored, 33, 40
  in series and parallel, 33, 37
  parallel-plate, 33, 36
  types of, 33, 40
Ceramic capacitor, 41
Charge, 1, 32
Coercive force, 49
Combination of waveforms, 82, 90
Composite magnetic circuits, 54
Conductance, 2
Contact potential, 121
Continuity tester, 141
Crest value, 81
C.R.O., 141
Cycle, 80

Damping, 138
DC circuit theory, 8
Depletion layer, 121
Dielectric, 32
Dielectric strength, 33
Digital instrument, 138

Electric field strength, 32, 33
  flux density, 32, 33
  force, 32

Electrolytic capacitor, 42
Electromagnetic induction, 61
Electrostatics, 32
Electronic voltmeter, 142
Energy, 1
  stored in capacitor, 33, 40
    inductor, 64, 72
Errors in measurement, 143, 148

Farad, 32
Faraday's laws, 61
Fleming.s left hand rule, 63
  right hand rule, 61
Flux density, magnetic, 48
Flux, magnetic, 48
Form factor, 81
Force, 1
  on a conductor, 62, 64
Frequency, 80, 84

Generator, 61, 79

Henry, 63
Hertz, 80
Hole, 120
Hysteresis, 49
  loop, 49

Impedance, 101
  triangle, 102
Induced emf, 62, 66
Inductance,
  self, 63, 70
  mutual, 63, 71
Inductive reactance, 100, 104
Instantaneous values, 81
Internal resistance, 9, 15
Instruments, comparison of, 139

Kirchhoff's laws, 9, 21

Lenz's law, 61

Magnetic circuits, 48
  composite, 54
  field strength, 48

155

Magnetic circuits (*cont.*)
    flux, 48
    flux density, 48
    force, 48
Magnetisation curves, 49, 50
Magnetising force, 48
Magnetomotive force, 48
Maximum value, 81
Mean value, 81
Measurements, 138
Measurement errors, 143, 148
Megger, 141
Mica capacitor, 41
Minority carriers, 121
Motor, 62
Moving coil instrument, 139, 143
    rectifier type instrument, 139
Moving iron instrument, 139, 144
Multimeter, 141
Multiplier, 140
Mutual inductance, 63

n-type material, 119

Ohmmeter, 140
Ohm's law, 8, 10

Paper capacitor, 41
Parallel-plate capacitor, 33, 36
Peak factor, 81
Peak to peak value, 81
Peak value, 81
Periodic time, 80, 84
Permeability,
    absolute, 49
    of free space, 48
    relative, 48
Permittivity, 32
Phasor, 81, 91
Plastic capacitor, 42
p–n junction, 121
Potential difference, 2
Potentiometer, 8, 143
Power, 1, 2, 83, 103, 110
    triangle, 104
Power factor, 104
p-type material, 119

Q-factor, 103
Quantity of electricity, 1

Reactance, 100, 101
Reactive power, 104
Rectification, 83, 84
Relative permeability, 48
    permittivity, 32
Reluctance, 49
Remanence, 49
Resistance, 2
    internal, 9, 15
Resistors in series and parallel, 8, 10
Resistivity, 118
Resonance, series, 103, 108
Root mean square value, 81

Screw rule, 62
Semiconductor diodes, 118, 122
Series-parallel circuits, 8, 12
Series resonance, 103, 108
Shunt, 140
SI units, 1, 2, 3
Sine waves, 80
    a.c. values of, 81, 87
    combination of, 82, 90
Superposition theorem, 9, 17

Tesla, 48
Transformer, 64, 71
Transistors, 128
    common base, 132
    common emitter, 132

Units, 1, 2, 3
Universal instruments, 141

Valency electrons, 119
Variable air capacitor, 40
Voltage gradient, 32
    magnification, 103
    triangle, 102
Voltmeter, 140
    electronic, 142

Wattmeter, 141
Waveforms, 80
Weber, 48
Wheatstone bridge, 142, 147
Work, 1

Zener effect, 124